Praise for past reports by the
Nongovernmental International Panel on Climate Change

Climate Change Reconsidered is a comprehensive, multidisciplinary compilation of technical papers covering a very large variety of important topics that will be appreciated by all who desire reliable, up-to-date information.

> — Larry Bell, endowed professor and director
> Sasakawa International Center for Space
> Architecture at the University of Houston

Many will treat *Climate Change Reconsidered* as a highly authoritative source of reference. It is in particular a standing rebuke to all those alarmists who deny the existence of hard science supporting the sceptical case. ... Given the increasing realisation that climate mitigation efforts are creating an economic crisis, and increasing popular scepticism about the alarmist scenario, this is a timely publication, and a key resource for all of us who are arguing for common sense.

> — Roger Helmer
> Member of the European Parliament

The 2011 edition of *Climate Change Reconsidered* is a quite extraordinary achievement. It should put to rest once and for all any notion that "the science is settled" on the subject of global warming, or that humanity and our planet face an imminent manmade climate change disaster.

> — Paul Driessen
> Author, *Eco-Imperialism*

I fully support the efforts of the Nongovernmental International Panel on Climate Change (NIPCC) and publication of its latest report, *Climate Change Reconsidered II: Physical Science*, to help the general public to understand the reality of global climate change.

> — Kumar Raina
> Former Deputy Director General
> Geological Survey of India

I've been waiting for this book for twenty years. It was a long wait, but I'm not disappointed. *Climate Change Reconsidered* is a *tour de force*.

> — E. Calvin Beisner, Ph.D.
> National Spokesman, Cornwall Alliance for the
> Stewardship of Creation

Highly informative, *Climate Change Reconsidered* ought to be required reading for scientists, journalists, policymakers, teachers, and students. It is an eye-opening read for everyone else (concerned citizens, taxpayers, etc.).

> — William Mellberg
> Author, *Moon Missions*

[T]here are several chapters in the NIPCC report that are substantially more thorough and comprehensive than the IPCC treatment, including 5 (Solar variability and climate cycles), 7 (Biological effects of carbon dioxide enrichment), 8 (Species extinction) and 9 (Human health effects). Further, the NIPCC's regional approach to analyzing extreme events and historical and paleo records of temperature, rainfall, streamflow, glaciers, sea ice, and sea-level rise is commendable and frankly more informative than the global analyses provided by the IPCC.

> — Dr. Judith Curry, professor and chair
> School of Earth and Atmospheric Sciences
> Georgia Institute of Technology

NIPCC's CCR-II report should open the eyes of world leaders who have fallen prey to the scandalous climate dictates by the IPCC. People are already suffering the consequences of sub-prime financial instruments. Let them not suffer more from IPCC's sub-prime climate science and models. That is the stark message of the NIPCC's CCR-II report.

> — M.I. Bhat, formerly professor and head
> Department of Geology and Geophysics
> University of Kashmir, India

Climate Change Reconsidered is a comprehensive, authoritative, and definitive reply to the IPCC reports.

> — Dr. Gerrit van der Lingen
> Christchurch, New Zealand

I was glad to see that a new report was coming from the NIPCC. The work of this group of scientists to present the evidence for natural climate warming and climate change is an essential counter-balance to the biased reporting of the IPCC. They have brought to focus a range of peer-reviewed publications showing that natural forces have in the past and continue today to dominate the climate signal. Considering the recent evidence that climate models have failed to predict the flattening of the global temperature curve, and that global warming seems to have ended some 15 years ago, the work of the NIPCC is particularly important.

— Ian Clark, professor, Department of Earth Sciences
University of Ottawa, Canada

Library shelves are cluttered with books on global warming. The problem is identifying which ones are worth reading. The NIPCC's CCR-II report is one of these. Its coverage of the topic is comprehensive without being superficial. It sorts through conflicting claims made by scientists and highlights mounting evidence that climate sensitivity to carbon dioxide increase is lower than climate models have until now assumed.

— Chris de Freitas, School of Environment
The University of Auckland, New Zealand

The CCR-II report correctly explains that most of the reports on global warming and its impacts on sea-level rise, ice melts, glacial retreats, impact on crop production, extreme weather events, rainfall changes, etc. have not properly considered factors such as physical impacts of human activities, natural variability in climate, lopsided models used in the prediction of production estimates, etc. There is a need to look into these phenomena at local and regional scales before sensationalization of global warming-related studies.

— S. Jeevananda Reddy
Former Chief Technical Advisor
United Nations World Meteorological Organization

Climate Change Reconsidered is simply the most comprehensive documentation of the case against climate alarmism ever produced. Basing policy on the scientifically incomplete and internally inconsistent reports of the UN's Intergovernmental Panel on Climate Change is no longer controversial – *Climate Change Reconsidered* shows that it is absolutely foolhardy, and anyone doing so is risking humiliation. It is a must-read for anyone who is accountable to the public, and it needs to be taken very, very seriously.

— Patrick J. Michaels, Director
Center for the Study of Science, Cato Institute

The claim by the UN IPCC that "global sea level is rising at an enhanced rate and swamping tropical coral atolls" does NOT agree with observational facts, and must hence be discarded as a serious disinformation. This is well taken in the CCR-II report.

— Nils-Axel Mörner, emeritus professor
Paleogeophysics & Geodynamics
Stockholm University, Sweden

CCR-II provides scientists, policy makers and other interested parties information related to the current state of knowledge in atmospheric studies. Rather than coming from a pre-determined politicized position that is typical of the IPCC, the NIPCC constrains itself to the scientific process so as to provide objective information. If we (scientists) are honest, we understand that the study of atmospheric processes/dynamics is in its infancy. Consequently, the work of the NIPCC and its most recent report is very important. It is time to move away from politicized science back to science – this is what NIPCC is demonstrating by example.

— Bruce Borders, professor of Forest Biometrics
Warnell School of Forestry and Natural Resources
University of Georgia

Why Scientists Disagree About Global Warming

The NIPCC Report on Scientific Consensus

Craig D. Idso, Robert M. Carter, S. Fred Singer

NIPCC

NONGOVERNMENTAL INTERNATIONAL PANEL
ON CLIMATE CHANGE

Published for the
Nongovernmental International Panel on Climate Change (NIPCC)
by The Heartland Institute
3939 North Wilke Road
Arlington Heights, Illinois 60004
phone 312/377-4000
www.heartland.org

Additional copies of this book are available from
The Heartland Institute for the following prices:

1-10 copies	$14.95 per copy
11-50 copies	$12.95 per copy
51-100 copies	$10.95 per copy
101 or more	$8.95 per copy

Printed in the United States of America
ISBN-13 978-1-934791-57-8
ISBN-10 1-934791-57-1

Manufactured in the United States of America

Contents

Preface

The global warming debate is one of the most consequential public policy debates taking place in the world today. Billions of dollars have been spent in the name of preventing global warming or mitigating the human impact on Earth's climate. Governments are negotiating treaties that would require trillions of dollars more to be spent in the years ahead.

A frequent claim in the debate is that a "consensus" or even "overwhelming consensus" of scientists embrace the more alarming end of the spectrum of scientific projections of future climate change. Politicians including President Barack Obama and government agencies including the National Aeronautics and Space Administration (NASA) claim "97 percent of scientists agree" that climate change is both man-made and dangerous.

As the authors of this book explain, the claim of "scientific consensus" on the causes and consequences of climate change is without merit. There is no survey or study showing "consensus" on any of the most important scientific issues in the climate change debate. On the contrary, there is extensive evidence of scientific disagreement about many of the most important issues that must be resolved before the hypothesis of dangerous man-made global warming can be validated.

Other authors have refuted the claim of a "scientific consensus" about global warming. This book is different in that it comprehensively and specifically rebuts the surveys and studies used to support claims of a consensus. It then summarizes evidence showing disagreement, identifies four reasons why scientists disagree about global warming, and then provides a detailed survey of the physical science of global warming based on the authors' previous work.

This book is based on a chapter in a forthcoming much larger examination of the climate change debate to be titled *Climate Change Reconsidered II: Benefits and Costs of Fossil Fuels*. That volume will finish the three-volume *Climate Change Reconsidered II* series, totaling some 3,000 pages and reporting the findings of more than 4,000 peer-reviewed articles on climate change.

This book and the larger volume that will follow it are produced by the Nongovernmental International Panel on Climate Change (NIPCC), an international panel of scientists and scholars who came together to understand the causes and consequences of climate change. NIPCC has no formal attachment to or sponsorship from any government or government agency. It also receives no corporate funding for its activities.

NIPCC seeks to objectively analyze and interpret data and facts without conforming to any specific agenda. This organizational structure and purpose stand in contrast to those of the United Nations' Intergovernmental Panel on Climate Change (IPCC), which is government -sponsored, politically motivated, and predisposed to believing that dangerous human-related global warming is a problem in need of a UN solution.

This volume, like past NIPCC reports, is edited and published by the staff of The Heartland Institute, a national nonprofit research and educational organization newly relocated from Chicago to suburban Arlington Heights, Illinois. The authors wish to acknowledge and thank Joseph L. Bast and Diane C. Bast, Heartland's seemingly tireless editing duo, for their help in getting this chapter ready for release before the rest of the volume in which it will eventually appear.

Craig D. Idso, Ph.D.
Chairman
Center for the Study
of Carbon Dioxide
and Global Change
(USA)

Robert M. Carter, Ph.D.
Emeritus Fellow
Institute of Public Affairs
(Australia)

S. Fred Singer, Ph. D.
Chairman
Science and
Environmental Policy
Project (USA)

Key Findings

Key findings of this book include the following:

No Consensus

- The most important fact about climate science, often overlooked, is that scientists disagree about the environmental impacts of the combustion of fossil fuels on the global climate.

- The articles and surveys most commonly cited as showing support for a "scientific consensus" in favor of the catastrophic man-made global warming hypothesis are without exception methodologically flawed and often deliberately misleading.

- There is no survey or study showing "consensus" on the most important scientific issues in the climate change debate.

- Extensive survey data show deep disagreement among scientists on scientific issues that must be resolved before the man-made global warming hypothesis can be validated. Many prominent experts and probably most working scientists disagree with the claims made by the United Nations' Intergovernmental Panel on Climate Change (IPCC).

Why Scientists Disagree

- Climate is an interdisciplinary subject requiring insights from many fields of study. Very few scholars have mastery of more than one or two of these disciplines.

- Fundamental uncertainties arise from insufficient observational evidence, disagreements over how to interpret data, and how to set the parameters of models.

- IPCC, created to find and disseminate research finding a human impact on global climate, is not a credible source. It is agenda-driven, a political rather than scientific body, and some allege it is corrupt.

- Climate scientists, like all humans, can be biased. Origins of bias include careerism, grant-seeking, political views, and confirmation bias.

Scientific Method vs. Political Science

- The hypothesis implicit in all IPCC writings, though rarely explicitly stated, is that dangerous global warming is resulting, or will result, from human-related greenhouse gas emissions.

- The null hypothesis is that currently observed changes in global climate indices and the physical environment, as well as current changes in animal and plant characteristics, are the result of natural variability.

- In contradiction of the scientific method, IPCC assumes its implicit hypothesis is correct and that its only duty is to collect evidence and make plausible arguments in the hypothesis's favor.

Flawed Projections

- IPCC and virtually all the governments of the world depend on global climate models (GCMs) to forecast the effects of human-related greenhouse gas emissions on the climate.

- GCMs systematically over-estimate the sensitivity of climate to carbon dioxide (CO_2), many known forcings and feedbacks are poorly modeled, and modelers exclude forcings and feedbacks that run counter to their mission to find a human influence on climate.

- NIPCC estimates a doubling of CO_2 from pre-industrial levels (from 280 to 560 ppm) would likely produce a temperature forcing of 3.7 Wm^{-2} in the lower atmosphere, for about ~1°C of *prima facie* warming.

- Four specific forecasts made by GCMs have been falsified by real-world data from a wide variety of sources. In particular, there has been no global warming for some 18 years.

False Postulates

- Neither the rate nor the magnitude of the reported late twentieth century surface warming (1979–2000) lay outside normal natural variability.

- The late twentieth century warm peak was of no greater magnitude than previous peaks caused entirely by natural forcings and feedbacks.

- Historically, increases in atmospheric CO_2 follow increases in temperature, they did not precede them. Therefore, CO_2 levels could not have forced temperatures to rise.

- Solar forcings are not too small to explain twentieth century warming. In fact, their effect could be equal to or greater than the effect of CO_2 in the atmosphere.

- A warming of 2°C or more during the twenty-first century would probably not be harmful, on balance, because many areas of the world would benefit from or adjust to climate change.

Unreliable Circumstantial Evidence

- Melting of Arctic sea ice and polar icecaps is not occurring at "unnatural" rates and does not constitute evidence of a human impact on the climate.

- Best available data show sea-level rise is not accelerating. Local and regional sea levels continue to exhibit typical natural variability – in some places rising and in others falling.

- The link between warming and drought is weak, and by some measures drought decreased over the twentieth century. Changes in the hydrosphere of this type are regionally highly variable and show a closer correlation with multidecadal climate rhythmicity than they do with global temperature.

■ No convincing relationship has been established between warming over the past 100 years and increases in extreme weather events. Meteorological science suggests just the opposite: A warmer world will see milder weather patterns.

■ No evidence exists that current changes in Arctic permafrost are other than natural or are likely to cause a climate catastrophe by releasing methane into the atmosphere.

Policy Implications

■ Rather than rely exclusively on IPCC for scientific advice, policymakers should seek out advice from independent, nongovernment organizations and scientists who are free of financial and political conflicts of interest.

■ Individual nations should take charge of setting their own climate policies based upon the hazards that apply to their particular geography, geology, weather, and culture.

■ Rather than invest scarce world resources in a quixotic campaign based on politicized and unreliable science, world leaders would do well to turn their attention to the real problems their people and their planet face.

Introduction

Probably the most widely repeated claim in the debate over global warming is that "97 percent of scientists agree" that climate change is man-made and dangerous. This claim is not only false, but its presence in the debate is an insult to science.

As the size of recent reports by the alarmist Intergovernmental Panel on Climate Change (IPCC) and its skeptical counterpart, the Nongovernmental International Panel on Climate (NIPCC) suggest, climate science is a complex and highly technical subject, making simplistic claims about what "all" or "most" scientists believe necessarily misleading. Regrettably, this hasn't prevented various politicians and activists from proclaiming a "scientific consensus" or even "overwhelming scientific consensus" that human activities are responsible for observed climate changes in recent decades and could have "catastrophic" effects in the future.

The claim that "97 percent of scientists agree" appears on the websites of government agencies such as the U.S. National Aeronautics and Space Administration (NASA, 2015) and even respected scientific organizations such as the American Association for the Advancement of Science (AAAS, n.d.), yet such claims are either false or meaningless.

Chapter 1 debunks surveys and abstract-counting exercises that allege to have found a "scientific consensus" in favor of the man-made global warming hypothesis and reports surveys that found no consensus on the most important issues in the debate. Chapter 2 explains why scientists disagree, finding the sources of disagreement in the interdisciplinary character of the issue, fundamental uncertainties concerning climate science, the failure of IPCC to be an independent and reliable source of research on the subject, and bias among researchers.

Chapter 3 explains the scientific method and contrasts it with the methodology used by IPCC and appeals to the "precautionary principle." Chapter 4 describes flaws in how IPCC uses global climate models to make projections about present and future climate changes and reports the findings of superior models that foresee much less global warming and even cooling. Chapter 5 critiques five postulates or assumptions that underlie IPCC's work, and Chapter 6 critiques five key pieces of circumstantial evidence relied on by IPCC. Chapter 7 reports the policy implications of these findings, and a brief summary and conclusion end this book.

Chapters 1 and 2 are based on previously published work by Joseph Bast (Bast, 2010, 2012, 2013; Bast and Spencer, 2014) and have been considerably revised and expanded for this publication. Chapters 3 to 7 are based on the *Summary for Policymakers* of *Climate Change Reconsidered II: Physical Science,* an earlier volume in the same series as the present book produced by the Nongovernmental International Panel on Climate Change (NIPCC) (Idso, Carter, and Singer, 2014). Although brief, this summary of climate science is based on an exhaustive review of the scientific literature. Lead authors Craig D. Idso, Robert M. Carter, and S. Fred Singer worked with a team of some 50 scientists to produce a 1,200-page report that is comprehensive, objective, and faithful to the scientific method. It mirrors and rebuts IPCC's Working Group 1 and Working Group 2 contributions to IPCC's 2014 *Fifth Assessment Report,* or AR5 (IPCC, 2014). Like IPCC reports, NIPCC reports cite thousands of articles appearing in peer-reviewed science journals relevant to the subject of human-induced climate change.

NIPCC authors paid special attention to research that was either overlooked by IPCC or contains data, discussion, or implications arguing against IPCC's claim that dangerous global warming is resulting, or will result, from human-related greenhouse gas emissions. Most notably, NIPCC's authors say IPCC has exaggerated the amount of warming likely to occur if the concentration of atmospheric CO_2 were to double, and such warming as occurs is likely to be modest and cause no net harm to the global environment or to human well-being. The principal findings from *CCR-II: Physical Science* are summarized in Figure 1.

Figure 1
Summary of NIPCC's Findings on Physical Science

■ Atmospheric carbon dioxide (CO_2) is a mild greenhouse gas that exerts a diminishing warming effect as its concentration increases.

■ Doubling the concentration of atmospheric CO_2 from its pre-industrial level, in the absence of other forcings and feedbacks, would likely cause a warming of ~0.3 to 1.1°C, almost 50 percent of which must already have occurred.

■ A few tenths of a degree of additional warming, should it occur, would not represent a climate crisis.

■ Model outputs published in successive IPCC reports since 1990 project a doubling of CO_2 could cause warming of up to 6°C by 2100. Instead, global warming ceased around the end of the twentieth century and was followed (since 1997) by 16 years of stable temperature.

■ Over recent geological time, Earth's temperature has fluctuated naturally between about +4°C and -6°C with respect to twentieth century temperature. A warming of 2°C above today, should it occur, falls within the bounds of natural variability.

■ Though a future warming of 2°C would cause geographically varied ecological responses, no evidence exists that those changes would be net harmful to the global environment or to human well-being.

■ At the current level of ~400 ppm we still live in a CO_2-starved world. Atmospheric levels 15 times greater existed during the Cambrian Period (about 550 million years ago) without known adverse effects.

■ The overall warming since about 1860 corresponds to a recovery from the Little Ice Age modulated by natural multidecadal cycles driven by ocean-atmosphere oscillations, or by solar variations at the de Vries (~208 year) and Gleissberg (~80 year) and shorter periodicities.

- Earth has not warmed significantly for the past 18 years despite an 8 percent increase in atmospheric CO_2, which represents 34 percent of all extra CO_2 added to the atmosphere since the start of the industrial revolution.

- No close correlation exists between temperature variation over the past 150 years and human-related CO_2 emissions. The parallelism of temperature and CO_2 increase between about 1980 and 2000 AD could be due to chance and does not necessarily indicate causation.

- The causes of historic global warming remain uncertain, but significant correlations exist between climate patterning and multidecadal variation and solar activity over the past few hundred years.

- Forward projections of solar cyclicity imply the next few decades may be marked by global cooling rather than warming, despite continuing CO_2 emissions.

Source: Idso, C.D., Carter, R.M., Singer, S.F. 2013. Executive Summary, *Climate Change Reconsidered II: Physical Science.* Chicago, IL: The Heartland Institute.

References

AAAS, n.d. What we know. American Academy for the Advancement of Science. Website. http://whatweknow.aaas.org/get-the-facts/. Last viewed on October 30, 2015.

Bast, J.L. 2010. Analysis: New international survey of climate scientists. The Heartland Institute. Website. https://www.heartland.org/policy-documents/analysis-new-international-survey-climate-scientists. Last viewed on October 30, 2015.

Bast, J.L. 2012. The myth of the 98%. *Heartland Policy Brief.* Chicago, IL: The Heartland Institute (October 1).

Bast, J.L. 2013. AMS survey shows no consensus on global warming. *Heartland*

Policy Brief. Chicago, IL: The Heartland Institute (November 28).

Bast, J.L and Spencer, R. 2014. The myth of the climate change '97%.' *Wall Street Journal* (May 26).

Idso, C.D, Carter, R.M., and Singer, S.F. (Eds.) 2013. *Climate Change Reconsidered II: Physical Science.* Chicago, IL: The Heartland Institute.

IPCC 2014. Pachauri, R. and Meyer, L. (Eds). *Climate Change 2014: Synthesis Report.* Contribution of Working Groups I, II and III to the Fifth Assessment Report of the Intergovernmental Panel on Climate Change. Geneva, Switzerland.

NASA, 2015. Scientific consensus: Earth's climate is warming. National Aeronautics and Space Administration. Website. http://climate.nasa.gov/scientific-consensus/. Last viewed on October 30, 2015.

1

No Consensus

Key findings of this chapter include the following:

■ The most important fact about climate science, often overlooked, is that scientists disagree about the environmental impacts of the combustion of fossil fuels on the global climate.

■ The articles and surveys most commonly cited as showing support for a "scientific consensus" in favor of the catastrophic man-made global warming hypothesis are without exception methodologically flawed and often deliberately misleading.

■ There is no survey or study showing "consensus" on the most important scientific issues in the climate change debate.

■ Extensive survey data show deep disagreement among scientists on scientific issues that must be resolved before the man-made global warming hypothesis can be validated. Many prominent experts and probably most working scientists disagree with the claims made by the United Nations' Intergovernmental Panel on Climate Change (IPCC).

Why Debate Consensus?

Environmental activists and their allies in the media often characterize climate science as an "overwhelming consensus" in favor of a single view that is sometimes challenged by a tiny minority of scientists funded by the fossil fuel industry to "sow doubt" or otherwise emphasize the absence of certainty on key aspects of the debate (Hoggan and Littlemore, 2009; Oreskes and Conway, 2010; Mann, 2012; Prothero, 2013). This popular narrative grossly over-simplifies the issue while libeling scientists who question the alleged consensus (Cook, 2014). This section reveals scientists do, in fact, disagree on the causes and consequences of climate change.

In May 2014, Secretary of State John Kerry warned graduating students at Boston College of the "crippling consequences" of climate change. "Ninety-seven percent of the world's scientists tell us this is urgent," he added (Kerry, 2014). Three days earlier, President Obama tweeted that "Ninety-seven percent of scientists agree: #climate change is real, man-made and dangerous" (Obama, 2014). What is the basis of these claims?

The most influential statement of this alleged consensus appears in the *Summary for Policymakers* of the *Fifth Assessment Report* (AR5) from the Intergovernmental Panel on Climate Change (IPCC): "It is extremely likely (95%+ certainty) that more than half of the observed increase in global average surface temperature from 1951 to 2010 was caused by the anthropogenic increase in greenhouse gas concentrations and other anthropogenic forcings together. The best estimate of the human-induced contribution to warming is similar to the observed warming over this period" (IPCC, 2013, p. 17).

In a "synthesis report" produced the following year, IPCC went further, claiming "Continued emission of greenhouse gases will cause further warming and long-lasting changes in all components of the climate system, increasing the likelihood of severe, pervasive and irreversible impacts for people and ecosystems. Limiting climate change would require substantial and sustained reductions in greenhouse gas emissions which, together with adaptation, can limit climate change risks" (IPCC, 2014, p. 8). In that same report, IPCC expresses skepticism that even reducing emissions will make a difference: "Many aspects of climate change and associated impacts will continue for centuries, even if anthropogenic emissions of greenhouse gases are stopped. The risks of abrupt or irreversible changes increase as the magnitude of the warming increases" (p. 16).

The media uncritically reported IPCC's claims with headlines such as "New Climate Change Report Warns of Dire Consequences" (Howard, 2014) and "Panel's Warning on Climate Risk: Worst Is Yet to Come"

(Gillis, 2014).

What evidence is there for a "scientific consensus" on the causes and consequences of climate change? What do scientists really say? Any inquiry along these lines must begin by questioning the legitimacy of the question. Science does not advance by consensus or a show of hands. Disagreement is the rule and consensus is the exception in most academic disciplines. This is because science is a process leading to ever-greater certainty, necessarily implying that what is accepted as true today will likely not be accepted as true tomorrow. As Albert Einstein famously once said, "No amount of experimentation can ever prove me right; a single experiment can prove me wrong" (Einstein, 1996).

Still, claims of a "scientific consensus" cloud the current debate on climate change. Many people, scientists included, refuse to believe scientists and other experts, even scholars eminent in the field, simply because they are said to represent minority views in the science community. So what do the surveys and studies reveal?

References

Cook, R. 2014. Merchants of smear. *Heartland Policy Brief* (September). Chicago, IL: The Heartland Institute.

Einstein, A. 1996. Quoted in A. Calaprice, *The Quotable Einstein*. Princeton, MA: Princeton University Press, p. 224.

Gillis, J. 2014. Panel's warning on climate risk: Worst is yet to come. *New York Times* (March 31).

Hoggan, J. and Littlemore, R. 2009. *Climate Cover-Up: The Crusade to Deny Global Warming*. Vancouver, BC Canada: Greystone Books.

Howard, B.C. 2014. IPCC highlights risks of global warming and closing window of opportunity. *National Geographic* (March 31).

IPCC 2013. Summary for policymakers. In *Climate Change 2013: The Physical Science Basis*. Contribution of Working Group I to the Fifth Assessment Report of the Intergovernmental Panel on Climate Change. Stocker, T.F., Qin, D., Plattner, G.-K., Tignor, M., Allen, S.K., Boschung, J., Nauels, A., Xia, Y., Bex, V., and Midgley, P.M. (Eds.) New York, NY: Cambridge University Press.

IPCC 2014. *Climate Change 2014: Synthesis Report.* Contribution of Working Groups I, II and III to the Fifth Assessment Report of the Intergovernmental Panel on Climate Change. Pachauri, R. and Meyer, L. (Eds.). Geneva, Switzerland.

Kerry, J. 2014. Remarks at Boston College's 138th commencement ceremony (May 19). http://www.state.gov/secretary/remarks/2014/05/226291.htm.

Mann, M.E. 2012. *The Hockey Stick and the Climate Wars: Dispatches from the Front Lines.* New York, NY: Columbia University Press.

Obama, B. 2014. Twitter. https://twitter.com/barackobama/status/335089477296988160.

Oreskes, N. and Conway, E.M. 2010. *Merchants of Doubt: How a Handful of Scientists Obscured the Truth on Issues From Tobacco Smoke to Global Warming.* New York: NY: Bloomsbury Press.

Prothero, D.R. 2013. *Reality Check: How Science Deniers Threaten Our Future.* Bloomington, IN: Indiana University Press.

Flawed Surveys

Claims of a "scientific consensus" on the causes and consequences of climate rely on a handful of essays and reports that either survey scientists or count the number of articles published in peer-reviewed journals that appear to endorse the positions of IPCC. As this section reveals, these surveys and abstract-counting exercises are deeply flawed and do not prove what those who cite them claim.

Oreskes, 2004

The most frequently cited source for a "consensus of scientists" is a 2004 essay for the journal *Science* written by science historian Naomi Oreskes (Oreskes, 2004). Oreskes reported examining abstracts from 928 papers reported by the Institute for Scientific Information database published in scientific journals from 1993 and 2003, using the key words "global climate change." Although not a scientist, she concluded 75 percent of the abstracts either implicitly or explicitly supported IPCC's view that human activities were responsible for most of the observed warming over the previous 50 years while none directly dissented.

Oreskes' essay, which was not peer-reviewed, became the basis of a book, *Merchants of Doubt* (Oreskes and Conway, 2010), an academic career built on claiming that global warming "deniers" are a tiny minority within the scientific community, and even a movie based on her book released in 2015. Her claims were repeated in former Vice President Al Gore's movie, *An Inconvenient Truth,* and in his book with the same title (Gore, 2006).

It is now widely agreed Oreskes did not distinguish between articles that acknowledged or assumed some human impact on climate, however small, and articles that supported IPCC's more specific claim that human emissions are responsible for more than 50 percent of the global warming observed during the past 50 years. The abstracts often are silent on the matter, and Oreskes apparently made no effort to go beyond those abstracts. Her definition of consensus also is silent on whether man-made climate change is dangerous or benign, a rather important point in the debate.

Oreskes' literature review inexplicably overlooked hundreds of articles by prominent global warming skeptics including John Christy, Sherwood Idso, Richard Lindzen, and Patrick Michaels. More than 1,350 such articles (including articles published after Oreskes' study was completed) are now identified in an online bibliography (Popular Technology.net, 2014).

Oreskes' methodology was flawed by assuming a nonscientist could determine the findings of scientific research by quickly reading abstracts of published papers. Indeed, even trained climate scientists are unable to do so because abstracts routinely do not accurately reflect their articles' findings. According to In-Uck Park *et al.* in research published in *Nature* in 2014 (Park *et al.*, 2014), abstracts routinely overstate or exaggerate research findings and contain claims that are irrelevant to the underlying research. The authors found "a mismatch between the claims made in the abstracts, and the strength of evidence for those claims based on a neutral analysis of the data, consistent with the occurrence of herding." They note abstracts often are loaded with "keywords" to ensure they are picked up by search engines and thus cited by other researchers.

Oreskes' methodology is further flawed, as are all the other surveys and abstract-counting exercises discussed in this section, by surveying the opinions and writings of scientists and often nonscientists who may write about climate but are by no means experts on or even casually familiar with the science dealing with attribution – that is, attributing a specific climate effect (such as a temperature increase) to a specific cause (such as rising CO_2 levels). Most articles simply reference or assume to be true the claims of IPCC and then go on to address a different topic, such as the effect of ambient temperature on the life-cycle of frogs, say, or correlations between temperature and outbreaks of influenza. Attribution is the issue the surveys ask about, but they ask people who have never studied the issue. The number of scientists actually knowledgeable about this aspect of the debate may be fewer than 100 in the world. Several are prominent skeptics (John Christy, Richard Lindzen, Patrick Michaels, and Roy Spencer, to name only four) and many others may be.

Monckton (2007) finds numerous other errors in Oreskes' essay including her use of the search term "global climate change" instead of "climate change," which resulted in her finding fewer than one-thirteenth of the estimated corpus of scientific papers on climate change over the stated period. Monckton also points out Oreskes never stated how many of the 928 abstracts she reviewed actually endorsed her limited definition of "consensus."

Medical researcher Klaus-Martin Schulte used the same database and search terms as Oreskes to examine papers published from 2004 to February 2007 and found fewer than half endorsed the "consensus" and only 7 percent did so explicitly (Schulte, 2008). His study is described in more detail below.

References

Gore, A. 2006. *An Inconvenient Truth: The Planetary Emergency of Global Warming and What We Can Do About It.* Emmaus, PA: Rodale Press.

Monckton, C. 2007. *Consensus? What consensus? Among climate scientists, the debate is not over.* Washington, DC: Science and Public Policy Institute.

Oreskes, N. 2004. Beyond the ivory tower: the scientific consensus on climate change. *Science* **306**: 5702 (December) 1686. DOI: 10.1126/science.1103618.

Oreskes, N. and Conway, E.M. 2010. *Merchants of Doubt: How a Handful of Scientists Obscured the Truth on Issues From Tobacco Smoke to Global Warming.* New York: NY: Bloomsbury Press.

Park, I.-U., Peacey, M.W., and Munafo, M.R. 2014. Modelling the effects of subjective and objective decision making in scientific peer review. *Nature* **506:** 93–96.

Popular Technology.net. 2014. 1350+ peer-reviewed papers supporting skeptic arguments against ACC/AGW alarmism. Website (February 12). http://www.populartechnology.net/2009/10/peer-reviewed-papers-supporting.html. Last viewed on September 23, 2015.

Schulte, K-M. 2008. Scientific consensus on climate change? *Energy & Environment* **19:** 2.

Doran and Zimmerman, 2009

In 2009, a paper by Maggie Kendall Zimmerman, at the time a student at the University of Illinois, and her master's thesis advisor Peter Doran was published in *EOS*. They claimed "97 percent of climate scientists agree" that mean global temperatures have risen since before the 1800s and that humans are a significant contributing factor (Doran and Zimmerman, 2009). This study, too, has been debunked.

The researchers sent a two-minute online survey to 10,257 Earth scientists working for universities and government research agencies, generating responses from 3,146 people. Solomon (2010) observed, "The two researchers started by altogether excluding from their survey the thousands of scientists most likely to think that the Sun, or planetary movements, might have something to do with climate on Earth – out were the solar scientists, space scientists, cosmologists, physicists, meteorologists and astronomers. That left the 10,257 scientists in disciplines like geology, oceanography, paleontology, and geochemistry that were somehow deemed more worthy of being included in the consensus. The two researchers also decided that scientific accomplishment should not be a factor in who could answer – those surveyed were determined by their place of employment (an academic or a governmental institution). Neither was academic qualification a factor – about 1,000 of those surveyed did not have a Ph.D., some didn't even have a master's diploma." Only 5 percent of respondents self-identified as climate scientists.

Even worse than the sample size, the bias shown in its selection, and the low response rate, though, is the irrelevance of the questions asked in the survey to the debate taking place about climate change. The survey asked two questions:

"Q1. When compared with pre-1800s levels, do you think that mean global temperatures have generally risen, fallen, or remained relatively constant?

Q2. Do you think human activity is a significant contributing factor in changing mean global temperatures?"

Overall, 90 percent of respondents answered "risen" to question 1 and 82 percent answered "yes" to question 2. The authors get their fraudulent "97 percent of climate scientists believe" sound bite by focusing on only 79 scientists who responded and "listed climate science as their area of expertise and who also have published more than 50 percent of their recent peer-reviewed papers on the subject of climate change."

Most skeptics of man-made global warming would answer those two questions the same way as alarmists would. At issue is not whether the climate warmed since the Little Ice Age or whether there is a human impact on climate, but whether the warming is unusual in rate or magnitude; whether that part of it attributable to human causes is likely to be beneficial or harmful on net and by how much; and whether the benefits of reducing human carbon dioxide emissions – i.e., reducing the use of fossil fuels – would outweigh the costs, so as to justify public policies aimed at reducing those emissions. The survey is silent on these questions.

The survey by Doran and Zimmerman fails to produce evidence that would back up claims of a "scientific consensus" about the causes or consequences of climate change. They simply asked the wrong people the wrong questions. The "98 percent" figure so often attributed to their survey refers to the opinions of only 79 climate scientists, hardly a representative sample of scientific opinion.

References

Doran, P.T. and Zimmerman, M.K. 2009. Examining the scientific consensus on climate change. *EOS* **90:** 3, 22–23. DOI: 10.1029/2009EO030002.

Solomon, L. 2010. 75 climate scientists think humans contribute to global warming. *National Post.* December 30.

Anderegg *et al.*, 2010

William R. Love Anderegg, then a student at Stanford University, used Google Scholar to identify the views of the most prolific writers on climate change. He claimed to find "(i) 97–98% of the climate researchers most actively publishing in the field support the tenets of ACC [anthropogenic climate change] outlined by the Intergovernmental Panel on Climate Change, and (ii) the relative climate expertise and scientific prominence of the researchers unconvinced of ACC are substantially below that of the convinced researchers" (Anderegg *et al.*, 2010). This college paper was published in *Proceedings of the National Academy of Sciences*, thanks to the addition of three academics as coauthors.

This is not a survey of scientists, whether "all scientists" or specifically climate scientists. Instead, Anderegg simply counted the number of articles he found on the Internet published in academic journals by 908 scientists. This counting exercise is the same flawed methodology utilized by Oreskes, falsely assuming abstracts of papers accurately reflect their findings. Further, Anderegg did not determine how many of these authors believe global warming is harmful or that the science is sufficiently established to be the basis for public policy. Anyone who cites this study in defense of these views is mistaken.

Anderegg *et al.* also didn't count as "skeptics" the scientists whose work exposes gaps in the man-made global warming theory or contradicts claims that climate change will be catastrophic. Avery (2007) identified several hundred scientists who fall into this category, even though some profess to "believe" in global warming.

Looking past the flashy "97–98%" claim, Anderegg *et al.* found the average skeptic has been published about half as frequently as the average alarmist (60 versus 119 articles). Most of this difference was driven by the hyper-productivity of a handful of alarmist climate scientists: The 50 most prolific alarmists were published an average of 408 times, versus only 89 times for the skeptics. The extraordinary publication rate of alarmists should raise a red flag. It is unlikely these scientists actually participated in most of the experiments or research contained in articles bearing their names.

The difference in productivity between alarmists and skeptics can be explained by several factors that do not include merit:

- Publication bias – articles that "find something," such as a statistically significant correlation that might suggest causation, are much more likely to get published than those that do not;

■ Heavy government funding of the search for one result but little or no funding for other results – the U.S. government alone paid $64 billion to climate researchers during the four years from 2010 to 2013, virtually all of it explicitly assuming or intended to find a human impact on climate and virtually nothing on the possibility of natural causes of climate change (Butos and McQuade, 2015, Table 2, p. 178);

■ Resumé padding – it is increasingly common for academic articles on climate change to have multiple and even a dozen or more authors, inflating the number of times a researcher can claim to have been published (Hotz, 2015). Adding a previously published researcher's name to the work of more junior researchers helps ensure approval by peer reviewers (as was the case, ironically, with Anderegg *et al.*);

■ Differences in the age and academic status of global warming alarmists versus skeptics – climate scientists who are skeptics tend to be older and more are emeritus than their counterparts on the alarmist side; skeptics are under less pressure and often are simply less eager to publish.

So what, exactly, did Anderegg *et al.* discover? That a small clique of climate alarmists had their names added to hundreds of articles published in academic journals, something that probably would have been impossible or judged unethical just a decade or two ago. Anderegg *et al.* simply assert those "top 50" are more credible than scientists who publish less, but they make no effort to prove this and there is ample evidence they are not (Solomon, 2008). Once again, the authors did not ask if authors believe global warming is a serious problem or if science is sufficiently established to be the basis for public policy. Anyone who cites this study as evidence of scientific support for such views is misrepresenting the paper.

References

Anderegg, W.R.L., Prall, J.W., Harold, J., and Schneider, S.H. 2010. Expert credibility in climate change. *Proceedings of the National Academy of Sciences* **107**: 27. 12107–12109.

Avery, D.T. 2007. 500 scientists whose research contradicts man-made global warming scares. The Heartland Institute (September 14). https://www.heartland.org/policy-documents/500-scientists-whose-research-cont radicts-man-made-global-warming-scares. Last viewed on October 30, 2015.

Butos, W.N. and McQuade, T.J. 2015. Causes and consequences of the climate science boom. *The Independent Review* **20**: 2 (Fall), 165–196.

Hotz, R.L. 2015. How many scientists does it take to write a paper? Apparently thousands. *The Wall Street Journal* (August 10).

Solomon, L. 2008. *The Deniers: The World Renowned Scientists Who Stood Up Against Global Warming Hysteria, Political Persecution, and Fraud**And those who are too fearful to do so*. Washington, DC: Richard Vigilante Books.

Cook *et al.*, 2013

In 2013, a paper by John Cook, an Australia-based blogger, and some of his friends published in *Environmental Research Letters* claimed their review of the abstracts of peer-reviewed papers from 1991 to 2011 found 97 percent of those that stated a position explicitly or implicitly suggested human activity is responsible for some warming (Cook *et al.*, 2013). This exercise in abstract-counting doesn't support the alarmist claim that climate change is both man-made and dangerous, and it doesn't even support IPCC's claim that a majority of global warming in the twentieth century was man-made.

This study was quickly debunked by Legates *et al.* (2013) in a paper published in *Science & Education*. Legates *et al.* found "just 0.03 percent endorsement of the standard definition of consensus: that most warming since 1950 is anthropogenic." They found "only 41 papers – 0.3 percent of all 11,944 abstracts or 1.0 percent of the 4,014 expressing an opinion, and not 97.1 percent – had been found to endorse the standard or quantitative hypothesis."

Scientists whose work questions the consensus, including Craig Idso, Nils-Axel Mörner, Nicola Scafetta, and Nir J. Shaviv, protested that Cook misrepresented their work (Popular Technology.net, 2013).

Richard Tol, a lead author of the United Nations' IPCC reports, said of the Cook report, "the sample of papers does not represent the literature. That is, the main finding of the paper is incorrect, invalid and unrepresentative" (Tol, 2013). On a blog of *The Guardian,* a British newspaper that had reported on the Cook report, Tol explained: "Cook's sample is not representative. Any conclusion they draw is not about 'the literature' but rather about the papers they happened to find. Most of the papers they studied are not about climate change and its causes, but many were taken as evidence nonetheless. Papers on carbon taxes naturally assume that carbon dioxide emissions cause global warming – but assumptions are not conclusions. Cook's claim of an increasing consensus over time is entirely due to an increase of the number of irrelevant papers that Cook and Co. mistook for evidence" (Tol, 2014).

Montford (2013) produced a blistering critique of Cook *et al.* in a report produced for the Global Warming Policy Foundation. He reveals the authors were marketing the expected results of the paper before the research itself was conducted; changed the definition of an endorsement of the global warming hypothesis mid-stream when it became apparent the abstracts they were reviewing did not support their original (IPCC-based) definition; and gave guidance to the volunteers recruited to read and score abstracts "suggest[ing] that an abstract containing the words 'Emissions of a broad range of greenhouse gases of varying lifetimes contribute to global climate change' should be taken as explicit but unquantified endorsement of the consensus. Clearly the phrase quoted could imply any level of human contribution to warming." Montford concludes "the consensus referred to is trivial" since the paper "said nothing about global warming being dangerous" and that "the project was not a scientific investigation to determine the extent of agreement on global warming, but a public relations exercise."

A group of Canadian retired Earth and atmospheric scientists called Friends of Science produced a report in 2014 that reviewed the four surveys and abstract-counting exercises summarized above (Friends of Science, 2014). The scientists searched the papers for the percentage of respondents or abstracts that explicitly agree with IPCC's declaration that human activity is responsible for more than half of observed warming. They found Oreskes found only 1.2 percent agreement; Doran and Zimmerman, 3.4 percent; Anderegg *et al.*, 66 percent; and Cook *et al.*, 0.54 percent. They conclude, "The purpose of the 97% claim lies in the psychological sciences, not in climate science. A 97% consensus claim is merely a 'social proof' – a powerful psychological motivator intended to make the public comply with the herd; to not be the 'odd man out.' Friends of Science deconstruction of these surveys shows there is no 97% consensus on human-caused global warming as claimed in these studies. None of these studies indicate any agreement with a catastrophic view of human-caused global warming" (p. 4).

References

Cook, J. Nuccitelli, D., Green, S.A., Richardson, M., Winkler, B., Painting, R., Way, R. Jacobs, P., and Skuce, A. 2013. Quantifying the consensus on anthropogenic global warming in the scientific literature. *Environmental Research Letters* **8:** 2.

Friends of Science. 2014. 97 Percent Consensus? No! Global Warming Math Myths & Social Proofs. Calgary, Canada: Friends of Science Society.

Legates, D.R., Soon, W., Briggs, W.M., and Monckton, C. 2013. Climate consensus and 'misinformation': A rejoinder to agnotology, scientific consensus, and the teaching and learning of climate change. *Science & Education* **24:** 3. 299–318.

Montford, A. 2013. Consensus? What consensus? *GWPF Note 5*. London, UK: Global Warming Policy Foundation.

Popular Technology.net. 2013. 97% Study falsely classifies scientists' papers, according to the scientists that published them. Website (May 21). http://www.populartechnology.net/2013/05/97-study-falsely-classifies-scientists. html. Last viewed on September 23, 2015.

Tol, R. 2013. Open letter to Professor Peter Høj, president and vice-chancellor, University of Queensland (August 2013). http://joannenova.com.au/2013/08/richard-tol-half-cooks-data-still-hidden-rest-s hows-result-is-incorrect-invalid-unrepresentative/.

Tol, R. 2014. The claim of a 97% consensus on global warming does not stand up. *The Guardian.* Blog (June 6). http://www.theguardian.com/environment/blog/2014/jun/06/97-consensus-global -warming. Last viewed on October 30, 2015.

Evidence of Lack of Consensus

In contrast to the studies described above, which try but fail to find a consensus in support of the claim that global warming is man-made and dangerous, many authors and surveys have found widespread disagreement or even that a majority of scientists oppose the alleged consensus. These surveys and studies generally suffer the same methodological errors as afflict the ones described above, but they suggest that even playing by the alarmists' rules, the results demonstrate disagreement rather than consensus.

Klaus-Martin Schulte, 2008

Schulte (2008), a practicing physician, observed, "Recently, patients alarmed by the tone of media reports and political speeches on climate change have been voicing distress, for fear of the imagined consequences of anthropogenic 'global warming.'" Concern that his patients were experiencing unnecessary stress "prompted me to review the literature available on 'climate change and health' via PubMed (http://www.ncbi.nlm.nih.gov/sites/entrez)" and then to attempt to replicate Oreskes' 2004 report.

"In the present study," Schulte wrote, "Oreskes' research was brought up to date by using the same search term on the same database to identify abstracts of 539 scientific papers published between 2004 and mid-February 2007." According to Schulte, "The results show a tripling of the mean annual publication rate for papers using the search term 'global climate change', and, at the same time, a significant movement of scientific opinion away from the apparently unanimous consensus which Oreskes had found in the learned journals from 1993 to 2003. Remarkably, the proportion of papers explicitly or implicitly rejecting the consensus has risen from zero in the period 1993–2003 to almost 6% since 2004. Six papers reject the consensus outright."

Schulte also found "Though Oreskes did not state how many of the papers she reviewed explicitly endorsed the consensus that human greenhouse-gas emissions are responsible for more than half of the past 50 years' warming, only 7% of the more recent papers reviewed here were explicit in endorsing the consensus even in the strictly limited sense she had defined. The proportion of papers that now explicitly or implicitly endorse the consensus has fallen from 75% to 45%."

Schulte's findings demonstrate that if Oreskes' methodology were correct and her findings for the period 1993 to 2003 accurate, then scientific publications in the more recent period of 2004–2007 show a strong tendency away from the consensus Oreskes claimed to have found. We can doubt the utility of the methodology used by both Oreskes and Schulte but recognize that the same methodology applied during two time periods reveals a significant shift from consensus to open debate on the causes of climate change.

Reference

Schulte, K-M. 2008. Scientific consensus on climate change? *Energy & Environment* **19:** 2.

Dennis Bray and Hans von Storch, 1996, 2003, 2008, 2010

Surveys by German scientists Dennis Bray and Hans von Storch conducted in 1996, 2003, 2008, and 2010 consistently found climate scientists have deep doubts about the reliability of the science underlying claims of man-made climate change (Bray and von Storch, 2007; Bray and von Storch, 2008; Bray and von Storch, 2010). This finding is seldom reported because the authors repeatedly portray their findings as supporting, as Bray wrote in 2010, "three dimensions of consensus, as it pertains to climate change science: 1. manifestation, 2. attribution, and 3. legitimation" (Bray, 2010). They do not.

One question in Bray and von Storch's latest survey (2010) asked scientists to grade, on a scale from 1 = "very inadequate" to 7 = "very adequate," the "data availability for climate change analysis." On this very important question, more respondents said "very inadequate" (1 or 2) than "very adequate" (6 or 7), with most responses ranging between 3 and 5.

Bray and von Storch summarized their survey results using a series of graphs plotting responses to each question. In their latest survey, 54 graphs show responses to questions addressing scientific issues as opposed to opinions about IPCC, where journalists tend to get their information, personal identification with environmental causes, etc. About a third show more skepticism than confidence, a third show more confidence than skepticism, and a third suggest equal amounts of skepticism and confidence.

For example, more scientists said "very inadequate" (1 or 2) than "very adequate" (6 or 7) when asked "How well do atmospheric models deal with the influence of clouds?" and "How well do atmospheric models deal with precipitation?" and "How well do atmospheric models deal with atmospheric convection?" and "The ability of global climate models to model sea-level rise for the next 50 years" and "The ability of global climate models to model extreme events for the next 10 years." These are not arcane or trivial matters in the climate debate.

Unfortunately, the Bray and von Storch surveys also show disagreement and outright skepticism about the underlying science of climate change don't prevent most scientists from expressing their opinion that man-made global warming is occurring and is a serious problem. On those questions, the distribution skews away from uncertainty and toward confidence. Observing this contradiction in their 1996 survey, Bray and von Storch described it as "an empirical example of 'postnormal science,'" the willingness to endorse a perceived consensus despite knowledge of contradictory scientific knowledge when the risks are perceived as being great (Bray and von Storch, 1999). Others might refer to this as cognitive dissonance, holding two contradictory opinions at the same time, or "herding," the well-documented tendency of academics facing uncertainty to ignore research that questions a perceived consensus position in order to advance their careers (Baddeleya, 2013).

On their face, Bray and von Storch's results should be easy to interpret. For at least a third of the questions asked, more scientists aren't satisfied than are with the quality of data, reliability of models, or predictions about future climate conditions. For another third, there is as much skepticism as there is strong confidence. Most scientists are somewhere in the middle, somewhat convinced that man-made climate change is occurring but concerned about lack of data and other fundamental uncertainties, far from the "95%+ certainty" claimed by IPCC.

Bray and von Storch are very coy in reporting and admitting the amount of disagreement their surveys find on the basic science of global warming, suggesting they have succumbed to the very cognitive dissonance they once described. But their data clearly reveal a truth: There is no scientific consensus.

References

Baddeleya, M. 2013. Herding, social influence and expert opinion. *Journal of Economic Methodology* **20:** 1. 35–44.

Bray, D. 2010. The scientific consensus of climate change revisited. *Environmental Science & Policy* **13:** 340–350.

Bray, D. and von Storch, H. 2007. The perspective of climate scientists on global climate change.' GKSS Report GKSS 2007/11. http://www.gkss.de/central_departments/library/publications/berichte_2007/inde x.html.en.

Bray, D. and von Storch, H. 2008. The perspectives of climate scientists on global climate change: A survey of opinions. http://coast.gkss.de/staff/storch/pdf/ CliSci2008.pdf.

Bray, D. and von Storch, H. 2010. A survey of climate scientists concerning climate science and climate change. http://www.academia.edu/2365610/The_Bray_and_von_Storch-survey_of_the_p erceptions_of_climate_scientists_2008_report_codebook_and_XLS_data.

Bray, D. and von Storch, H. 1999: Climate science: An empirical example of postnormal science. *Bulletin of the American Meteorological Society* **80:** 439–455.

Verheggen *et al.*, 2014, 2015

Verheggen *et al.* (2014) and Strengers, Verheggen, and Vringer (2015) reported the results of a survey they conducted in 2012 of contributors to IPCC reports, authors of articles appearing in scientific literature, and signers of petitions on global warming (but apparently not the Global Warming Petition Project, described below). By the authors' own admission, "signatories of public statements disapproving of mainstream climate science ... amounts to less than 5% of the total number of respondents," suggesting the sample is heavily biased toward pro-"consensus" views. Nevertheless, this survey found fewer than half of respondents agreed with IPCC's most recent claims.

A total of 7,555 authors were contacted and 1,868 questionnaires were returned, for a response rate of 29 percent. The authors asked specifically about agreement or disagreement with IPCC's claim in its *Fifth Assessment Report* (AR5) that it is "virtually certain" or "extremely likely" that net anthropogenic activities are responsible for more than half of the observed increase in global average temperatures in the past 50 years.

When asked "What fraction of global warming since the mid 20th century can be attributed to human induced increases in atmospheric greenhouse gas (GHG) concentrations?", 64 percent chose fractions of 51 percent or more, indicating agreement with IPCC AR5. (Strengers, Verheggen, and Vringer, 2015, Figure 1a.1) When those who chose fractions of 51 percent or more were asked, "What confidence level would you ascribe to your estimate that the anthropogenic GHG warming is more than 50%?", 65 percent said it was "virtually certain" or "extremely likely," the language used by IPCC to characterize its level of confidence (*Ibid.*, Figure 1b).

The math is pretty simple: Two-thirds of the authors in this survey – a sample heavily biased toward IPCC's point of view by including virtually all its editors and contributors – agreed with IPCC on the impact of human emissions on the climate, and two-thirds of those who agreed were as confident as IPCC in that finding. Sixty-five percent of 64 percent is 41.6 percent, so fewer than half of the survey's respondents support IPCC. More precisely – since some responses were difficult to interpret – 42.6 percent (797 of 1,868) of respondents were highly confident that more than 50 percent of the warming is human-caused.

This survey shows IPCC's position on global warming is the minority perspective in this part of the science community. Since the sample was heavily biased toward contributors to IPCC reports and academics most likely to publish, one can assume a survey of a larger universe of scientists would reveal even less support for IPCC's position.

Like Bray and von Storch (2010) discussed above, and Stenhouse *et al.*, (2014) discussed below, Verheggen *et al.* seem embarrassed by their findings and hide them in tables in a report issued a year after their original publication rather than explain them in the text of their peer-reviewed article. It took the efforts of a blogger to call attention to the real data (Fabius Maximus, 2015). Once again, the data reveal no scientific consensus.

References

Bray, D. and von Storch, H. 2010. A survey of climate scientists concerning climate science and climate change.
http://www.academia.edu/2365610/The_Bray_and_von_Storch-survey_of_the_p erceptions_of_climate_scientists_2008_report_codebook_and_XLS_data.

Fabius Maximus, 2015. Website. New study undercuts key IPCC finding.
http://fabiusmaximus.com/2015/07/29/new-study-undercuts-ipcc-keynote-findin g-87796/. Last viewed on September 24, 2015.

Strengers, B., Verheggen, B., and Vringer, K. 2015. Climate science survey questions and responses (April 10). PBL Netherlands Environmental Assessment Agency.
http://www.pbl.nl/sites/default/files/cms/publicaties/pbl-2015-climate-science-su rvey-questions-and-responses_01731.pdf.

Stenhouse, N., Maibach, E., Cobb, S., Ban, R., Bleistein, A., Croft, P., Bierly, E., Seitter, K., Rasmussen, G., and Leiserowitz, A. 2014: Meteorologists' views about global warming: A survey of American Meteorological Society professional members. *Bulletin of the American Meteorological Society* **95**: 1029–1040.

Verheggen, B., Strengers, B., Cook, J. van Dorland, R., Vringer, K., Peters, J. Visser, H., and Meyer, L. 2014. Scientists' views about attribution of global warming. *Environmental Science & Technology* **48**: 16. 8963–8971, DOI: 10.1021/es501998e

Surveys of Meteorologists and Environmental Professionals

The American Meteorological Society (AMS) reported in 2013 that only 52 percent of AMS members who responded to its survey reported believing the warming of the past 150 years was man-made (Stenhouse *et al.*, 2014). The finding was reported in a table on the last page of the pre-publication version of the paper and was not even mentioned in the body of the peer-reviewed article.

From an earlier publication of the survey's results (Maibach *et al.*, 2012) it appears 76 percent of those who believe in man-made global warming also believe it is "very harmful" or "somewhat harmful," so it appears 39.5 percent of AMS members responding to the survey say they believe man-made global warming could be dangerous. Once again, this finding doesn't appear in the peer-reviewed article.

Questions asked in the AMS survey reveal political ideology is the strongest or second strongest factor in determining a scientist's position on global warming. But the published report doesn't reveal whether all or just nearly all of the AMS members who believe man-made global warming is dangerous self-identify as being liberals. In light of the numbers presented above, this appears likely.

Other surveys of meteorologists also found a majority oppose the alleged consensus (Taylor, 2010a, 2010b). A 2006 survey of scientists in the U.S. conducted by the National Registry of Environmental Professionals, for example, found 41 percent disagreed the planet's recent warmth "can be, in large part, attributed to human activity," and 71 percent disagreed recent hurricane activity is significantly attributable to human activity (Taylor, 2007).

References

Maibach, E., Stenhouse, N., Cobb, S., Ban, R., Bleistein, A., *et al.* 2012. American Meteorological Society member survey on global warming: Preliminary findings (February 12). Fairfax, VA: Center for Climate Change Communication.

Stenhouse, N., Maibach, E., Cobb, S., Ban, R., Bleistein, A., Croft, P., Bierly, E., Seitter, K., Rasmussen, G., and Leiserowitz, A. 2014: Meteorologists' views about global warming: A survey of American Meteorological Society professional members. *Bulletin of the American Meteorological Society* **95:** 1029–1040.

Taylor, J.M. 2010a. Majority of broadcast meteorologists skeptical of global warming crisis. *Environment & Climate News* (April).

Taylor, J.M. 2010b. Meteorologists reject U.N.'s global warming claims. *Environment & Climate News* (February).

Taylor, J.M. 2007. Warming debate not over, survey of scientists shows. *Environment & Climate News* (February).

Global Warming Petition Project

The Global Warming Petition Project (2015) is a statement about the causes and consequences of climate change signed by 31,478 American scientists, including 9,021 with Ph.D.s. The full statement reads:

> We urge the United States government to reject the global warming agreement that was written in Kyoto, Japan in December, 1997, and any other similar proposals. The proposed limits on greenhouse gases would harm the environment, hinder the advance of science and technology, and damage the health and welfare of mankind.
>
> There is no convincing scientific evidence that human release of carbon dioxide, methane, or other greenhouse gases is causing or will, in the foreseeable future, cause catastrophic heating of the Earth's atmosphere and disruption of the Earth's climate. Moreover, there is substantial scientific evidence that increases in atmospheric carbon dioxide produce many beneficial effects upon the natural plant and animal environments of the Earth.

This is a remarkably strong statement of dissent from the perspective advanced by IPCC. The fact that more than ten times as many scientists have signed it as are alleged to have "participated" in some way or another in the research, writing, and review of IPCC's *Fourth Assessment Report* is very significant. These scientists actually endorse the statement that appears above. By contrast, fewer than 100 of the scientists (and nonscientists) who are listed in the appendices to IPCC reports actually participated in the writing of the all important *Summary for Policymakers* or the editing of the final report to comply with the summary, and therefore could be said to endorse the main findings of that report.

The Global Warming Petition Project has been criticized for including names of suspected nonscientists, including names submitted by environmental activists for the purpose of discrediting the petition. But the organizers of the project painstakingly reconfirmed the authenticity of the names in 2007, and a complete directory of those names appeared as an appendix to *Climate Change Reconsidered: Report of the Nongovernmental International Panel on Climate Change (NIPCC),* published in 2009 (Idso and Singer, 2009). For more information about The Petition Project, including the text of the letter endorsing it written by the late Dr. Frederick

Seitz, past president of the National Academy of Sciences and president emeritus of Rockefeller University, visit the project's Web site at www.petitionproject.org.

References

Global Warming Petition Project. 2015. Global warming petition project. Website. http://www.petitionproject.org/ Last viewed September 23, 2015.

Idso, C.D. and Singer, S.F. 2009. (Eds.) *Climate Change Reconsidered: Report of the Nongovernmental International Panel on Climate Change (NIPCC).* Chicago, IL: The Heartland Institute.

Admissions of Lack of Consensus

Even prominent "alarmists" in the climate change debate admit there is no consensus. Phil Jones, director of the Climatic Research Unit at the University of East Anglia, when asked if the debate on climate change is over, told the BBC, "I don't believe the vast majority of climate scientists think this. This is not my view" (BBC News, 2010). When asked, "Do you agree that according to the global temperature record used by IPCC, the rates of global warming from 1860–1880, 1910–1940 and 1975–1998 were identical?" Jones replied, "the warming rates for all 4 periods are similar and not statistically significantly different from each other." Finally, when asked "Do you agree that from 1995 to the present there has been no statistically-significant global warming" he answered "yes." Jones' replies to the second and third questions contradict claims made by IPCC.

Mike Hulme, also a professor at the University of East Anglia and a contributor to IPCC reports, wrote in 2009: "What is causing climate change? By how much is warming likely to accelerate? What level of warming is dangerous? – represent just three of a number of contested or uncertain areas of knowledge about climate change" (Hulme, 2009, p. 75). He admits "Uncertainty pervades scientific predictions about the future performance of global and regional climates. And uncertainties multiply when considering all the consequences that might follow from such changes in climate" (p. 83). On the subject of IPCC's credibility, he admits it is "governed by a Bureau consisting of selected governmental representatives, thus ensuring that the Panel's work was clearly seen to be serving the needs of government and policy. The Panel was not to be a self-governing body of independent scientists" (p. 95). All this is exactly what IPCC critics have been saying for years.

* * *

As this summary makes apparent, there is no survey or study that supports the claim of a scientific consensus that global warming is both man-made and a problem, and ample evidence to the contrary. There is no scientific consensus on global warming.

References

BBC News. 2010. Q&A: Professor Phil Jones (February 13). https://www.heartland.org/policy-documents/qa-professor-phil-jones.

Hulme, M. 2009. *Why We Disagree About Climate Change: Understanding Controversy, Inaction and Opportunity.* New York, NY: Cambridge University Press.

2

Why Scientists Disagree

Key findings in this section include the following:

- Climate is an interdisciplinary subject requiring insights from many fields. Very few scholars have mastery of more than one or two of these disciplines.

- Fundamental uncertainties arise from insufficient observational evidence, disagreements over how to interpret data, and how to set the parameters of models.

- The United Nations' Intergovernmental International Panel on Climate Change (IPCC), created to find and disseminate research finding a human impact on global climate, is not a credible source. It is agenda-driven, a political rather than scientific body, and some allege it is corrupt.

- Climate scientists, like all humans, can be biased. Origins of bias include careerism, grant-seeking, political views, and confirmation bias.

Conflict of Disciplines

Climate is an interdisciplinary subject requiring insights from many fields. Very few scholars have mastery of more than one or two of these disciplines.

One reason disagreement among those participating in the climate change debate may be sharper and sometimes more personal than is observed in debates on other topics is because climate is an interdisciplinary subject requiring insights from astronomy, biology, botany, cosmology, economics, geochemistry, geology, history, oceanography, paleontology, physics, and scientific forecasting and statistics, among other disciplines. Very few scholars in the field have mastery of more than one or two of these disciplines.

Richard S. Lindzen, an atmospheric physicist at MIT, observed, "Outside any given specialty, there are few – including scientists – who can distinguish one scientist from another, and this leaves a great deal of latitude for advocates and politicians to invent their own 'experts.' ... In effect, once political action is anticipated, the supporting scientific position is given a certain status whereby objections are reckoned to represent mere uncertainty, while scientific expertise is strongly discounted" (Lindzen, 1996, p. 98).

When an expert in one field, say physics, presents an estimate of the climate's sensitivity to rising carbon dioxide levels, an expert in another field, say biology, can quickly challenge his understanding of the carbon cycle, whereby huge volumes of carbon dioxide are added to and removed from the atmosphere. Unless the physicist is intimately familiar with the literature on the impact of rising levels of CO_2 on photosynthesis, plant growth, and carbon sequestration by plants and aquatic creatures, he or she is missing the bigger picture and is likely to be wrong. But so too will the biologist miss the "big picture" if he or she doesn't understand the transfer of energy at the top of the atmosphere and how the effects of CO_2 change logarithmically as its concentration rises.

Geologists view time in millennia and eons and are aware of huge fluctuations in both global temperatures and carbon dioxide concentrations in the atmosphere, with the two often moving in different directions. They scoff at physicists and botanists who express concern over a historically tiny increase in carbon dioxide concentrations of 100 parts per million and a half-degree C increase in temperature over the course of a century. But how many geologists understand the impact of even relatively small changes in temperature or humidity on the range and health of some plants and animals?

Economists are likely to ask if the benefits of trying to "stop" global warming outweigh the benefits of providing clean water or electricity to billions of people living in terrible poverty. If not, wouldn't it be wiser – better for humanity and perhaps even wildlife – to focus on helping people today become more prosperous and consequently more concerned about protecting the environment and able to afford to adapt to changes in weather regardless of their causes? But do economists properly value the contribution of ecological systems to human welfare, or apply properly the "discount rates" they use to measure costs and benefits that occur far in the future?

Simon (1999) observed another consequence of this tunnel vision. Scientists are often optimistic about the safety of the environment when it relates to subjects encompassing their own area of research and expertise, but are pessimistic about risks outside their range of expertise. Simon wrote:

> This phenomenon is apparent everywhere. Physicians know about the extraordinary progress in medicine that they fully expect to continue, but they can't believe in the same sort of progress in natural resources. Geologists know about the progress in natural resources that pushes down their prices, but they worry about food. Even worse, some of those who are most optimistic about their own areas point with alarm to other issues to promote their own initiatives. The motive is sometimes self-interest (pp. 47–8).

The climate change debate resembles the famous tale of a group of blind men touching various parts of an elephant, each arriving at a very different idea of what it is like: to one it is like a tree, to another, a snake, and to a third, a wall. A wise man tells the group, "You are all right. An elephant has all the features you mentioned." But how many physicists, geologists, biologists, and economists want to be told they are missing "the big picture" or that their earnest concern and good research aren't enough to describe a complex phenomenon, and therefore not a reliable guide to making decisions about what mankind should do? Few indeed.

This source of disagreement seems obvious but is seldom discussed. Scientists (both physical scientists and social scientists) make assertions and predictions claiming high degrees of "confidence," a term with precise meaning in science but turned into an empty tool of rhetoric by IPCC and its allies, that are wholly unjustified given their training and ignorance of large parts of the vast literature regarding climate.

References

Lindzen, R.S. 1996 Chapter 5. Science and politics: Global warming and eugenics. In Hahn, R.W. (Ed.) *Risks, Costs, and Lives Saved: Getting Better Results from Regulation.* New York, NY: Oxford University Press. 85–103.

Simon, J. 1999. *Hoodwinking the Nation.* New Brunswick, NJ: Transaction Publishers.

Scientific Uncertainties

Fundamental uncertainties arise from insufficient observational evidence, disagreements over how to interpret data, and how to set the parameters of models.

The claim that human activities are causing or will cause catastrophic global warming or climate is a rebuttable hypothesis, not a scientific theory and certainly not the "consensus" view of the science community. The human impact on climate remains a puzzle. As Bony *et al.* wrote in 2015, "Fundamental puzzles of climate science remain unsolved because of our limited understanding of how clouds, circulation and climate interact" (abstract).

Reporting in *Nature* on Bony's study, Quirin Schiermeier wrote, "There is a misconception that the major challenges in physical climate science are settled. 'That's absolutely not true,' says Sandrine Bony, a climate researcher at the Laboratory of Dynamic Meteorology in Paris. 'In fact, essential physical aspects of climate change are poorly understood'" (Schiermeier, 2015, p. 140). Schiermeier goes on to write, "large uncertainties persist in 'climate sensitivity,' the increase in average global temperature caused by a given rise in the concentration of carbon dioxide," citing Bjorn Stevens, a director at the Max Planck Institute for Meteorology in Hamburg, Germany (*Ibid.*). Bony has also identified uncertainty in climate science in the journal *Science* (Stevens and Bony, 2013).

The first two volumes in the *Climate Change Reconsidered II* series cited literally thousands of peer-reviewed articles and studies revealing the extensive uncertainty acknowledged by Bony *et al.* Since the *Summaries for Policymakers* of those volumes appear below (Sections 2.3 to 2.7 of this chapter and Chapter 5), there is no need to repeat them here. Instead, it is useful to ponder the views of two prominent climate scientists whose scientific contributions to the debate are widely acknowledged.

Richard S. Lindzen, quoted earlier, is one of the world's most distinguished atmospheric physicists. According to the biography on MIT's website, "he has developed models for the Earth's climate with specific concern for the stability of the ice caps, the sensitivity to increases in CO_2, the origin of the 100,000 year cycle in glaciation, and the maintenance of regional variations in climate. Prof. Lindzen is a recipient of the AMS's Meisinger, and Charney Awards, the AGU's Macelwane Medal, and the Leo Huss Walin Prize. He is a member of the National Academy of Sciences, and the Norwegian Academy of Sciences and Letters, and a fellow of the American Academy of Arts and Sciences, the American Association for the Advancement of Sciences, the American Geophysical Union and the American Meteorological Society.

Lindzen is a corresponding member of the NAS Committee on Human Rights, and has been a member of the NRC Board on Atmospheric Sciences and Climate and the Council of the AMS. He has also been a consultant to the Global Modeling and Simulation Group at NASA's Goddard Space Flight Center, and a Distinguished Visiting Scientist at California Institute of Technology's Jet Propulsion Laboratory." He received his Ph.D. from Harvard University in 1964.

According to Lindzen (1996), there are three principal areas of uncertainty in climate science:

- "First, the basic greenhouse process is not simple. In particular, it is not merely a matter of the bases that absorb heat radiation – greenhouse gases – keeping the earth warm. If it were, the natural greenhouse would be about four times more effective than it actually is. ...

- "Second, the most important greenhouse gas in the atmosphere is water vapor. ... Roughly speaking, changes in relative humidity on the order of 1.3 to 4 percent are equivalent to the effect of doubling carbon dioxide. Our measurement uncertainty for trends in water vapor is in excess of 10 percent, and once again, model errors are known to substantially exceed measurement errors in a very systematic way.

- "Third, the direct impact of doubling carbon dioxide on the earth's temperature is rather small: on the order of .3 degrees C. Larger predictions depend on positive feedbacks… [T]hose factors arise from models with errors in those factors."

"[T]here is very little argument about the above points," Lindzen wrote. "They are, for the most part, textbook material showing that there are errors and uncertainties in physical processes central to model predictions that are an order of magnitude greater than the climate forcing due to a putative doubling of carbon dioxide. There is, nonetheless, argument over whether the above points mean that the predicted significant response to increased carbon dioxide is without meaningful basis. Here there is disagreement" (pp. 86–7). For Lindzen's more recent views (which are similar) see Lindzen (2012).

A second recognized authority is Judith Curry, a professor and former chair of the School of Earth and Atmospheric Sciences at the Georgia Institute of Technology. Her Ph.D. in geophysical sciences is from the University of Chicago and she served for three decades on the faculties of the University of Wisconsin-Madison, Purdue, Penn State, University of Colorado-Boulder, and since 2002 at the Georgia Institute of Technology. She is an elected fellow of the American Geophysical Union and councilor and fellow of the American Meteorological Society.

Curry delivered a speech on June 15, 2015 to the British House of Lords. Titled "State of the climate debate in the U.S.," the prepared text of her remarks is available online (Curry, 2015). Curry wrote, "there is widespread agreement" on three basic tenets: "Surface temperatures have increased since 1880, humans are adding carbon dioxide to the atmosphere, [and] carbon dioxide and other greenhouse gases have a warming effect on the planet." However, she wrote, "there is disagreement about the most consequential issues," which she lists as the following:

- "Whether the warming since 1950 has been dominated by human causes
- "How much the planet will warm in the 21st century
- "Whether warming is 'dangerous'
- "Whether we can afford to radically reduce CO_2 emissions, and whether reduction will improve the climate"

Observing the "growing divergence between models and observations," she poses three questions:

- "Are climate models too sensitive to greenhouse forcing?
- "Is the modeled treatment of natural climate variability inadequate?
- "Are climate model projections of 21st century warming too high?"

After observing surveys show most scientists seem to accept IPCC's claims, she wrote, "Nevertheless, a great deal of uncertainty remains, and there is plenty of room for disagreement. So why do scientists disagree?" She gives five possible reasons:

- "Insufficient observational evidence
- "Disagreement about the value of different classes of evidence
- "Disagreement about the appropriate logical framework for linking and assessing the evidence
- "Assessments of areas of ambiguity & ignorance
- "And finally, the politicization of the science can torque the science in politically desired directions."

"None of the most consequential scientific uncertainties are going to be resolved any time soon," Curry wrote. "[T]here is a great deal of work still to do to understand climate change. And there is a growing realization that unpredictable natural climate variability is important."

All of this concurs with the findings of NIPCC and was documented at great length in *Climate Change Reconsidered II: Physical Science* and *Climate Change Reconsidered II: Biological Impacts* (Idso *et al.,* 2013; Idso *et al.,* 2014).

References

Bony, S., Stevens, B., Frierson, D.M.W., Jakob, C., Kageyama, M., Pincus, R., Shepherd, T.G., Sherwood, S.C., Siebesma, A.P., Sobel, A.H., Watanebe, M., and Webb, M.J. 2015. *Nature Geoscience* **8**: 261–268, doi:10.1038/ngeo2398.

Curry, J. 2015. State of the climate debate in the U.S. Remarks to the U.K. House of Lords, June 15. Climate Etc. Website. http://judithcurry.com/2015/06/15/state-of-the-climate-debate-in-the-u-s/.

Idso, C.D, Carter, R.M., and Singer, S.F. (Eds.) 2013. *Climate Change Reconsidered II: Physical Science.* Chicago, IL: The Heartland Institute.

Idso, C.D, Idso, S.B., Carter, R.M., and Singer, S.F. (Eds.) 2014. *Climate Change Reconsidered II: Biological Impacts*. Chicago, IL: The Heartland Institute.

Lindzen, R.S. 1996. Chapter 5. Science and politics: Global warming and eugenics. In Hahn, R.W. (Ed.) Risks, *Costs, and Lives Saved: Getting Better Results from Regulation*. New York, NY: Oxford University Press. 85–103.

Lindzen, R.S. 2012. Climate science: Is it currently designed to answer questions? *Euresis Journal* **2:** (Winter).

Schiermeier, Q. 2015. Physicists, your planet needs you. *Nature* **520:** (April 2).

Stevens, B. and Bony, S. 2013. What are climate models missing? *Science* **340:** 6136 (31 May), 1053-1054. DOI: 10.1126/science.1237554.

Failure of IPCC

The Intergovernmental Panel on Climate Change (IPCC), created to find and disseminate research finding a human impact on global climate, is not a credible source. It is agenda-driven, a political rather than scientific body, and some allege it is corrupt.

According to Bray (2010), "In terms of providing future projection[s] of the global climate, the most significant player in setting the agenda is the Intergovernmental Panel on Climate Change (IPCC). It is typically assumed that IPCC, consisting of some 2500 climate scientists, after weighing the evidence, arrived at a consensus that global temperatures are rising and the most plausible cause is anthropogenic in nature." As this section will explain, that trust is misplaced.

Prior to the mid-1980s very few climate scientists believed man-made climate change was a problem. This non-alarmist "consensus" on the causes and consequences of climate change included nearly all the leading climate scientists in the world, including Roger Revelle, often identified as one of the first scientists to "sound the alarm" over man-made global warming (Solomon, 2008; Singer, Revelle and Starr, 1992).

Most of the reports purporting to show a "consensus" beginning in the 1980s came from and continue to come from committees funded by government agencies tasked with finding a new problem to address or by liberal foundations with little or no scientific expertise (Darwall, 2013;

Carlin, 2015; Moore *et al.*, 2014). These committees, one of which was IPCC, often produced reports making increasingly bold and confident assertions about future climate impacts, but they invariably included statements admitting deep scientific uncertainty (Weart, 2015). Reports of IPCC, including drafts of the latest *Fifth Assessment Report,* are replete with examples of this pattern.

It is common for committees seeking consensus reports to include qualifications and admissions of uncertainty and even publish dissenting reports by committee members. This common practice had an unintended result in the climate debate. Politicians, environmental activists, and rent-seeking corporations in the renewable energy industry began to routinely quote IPCC's alarming claims and predictions shorn of the important qualifying statements expressing deep doubts and reservations. Rather than protest this mishandling of its work, IPCC encouraged it by producing Summaries for Policymakers that edit away or attempt to hide qualifying statements. IPCC news releases have become more and more alarmist over time until they are indistinguishable from the news releases and newsletters of environmental groups. In fact, many of those IPCC news releases were written or strongly influenced by professional environmental activists who had effectively taken over the organization.

Some climate scientists spoke out early and forcefully against this corruption of science (Idso, 1982; Landsberg, 1984; Idso, 1989; Singer, 1989; Jastrow, Nierenberg, and Seitz, 1990; Balling, 1992; Michaels, 1992) but their voices were difficult to hear amid a steady drumbeat of doomsday forecasts produced by environmentalists and their allies in the mainstream media.

Perhaps the most conspicuous and consequential example of this practice occurred in 2006 in the form of a movie titled *An Inconvenient Truth*, produced by former Vice President Al Gore, and Gore's book with the same title (Gore, 2006). The movie earned Gore a Nobel Peace Prize (shared with IPCC), yet it made so many unsubstantiated claims and over-the-top predictions it was declared "propaganda" by a UK judge and schools there were ordered to give students a study guide identifying and correcting its errors before showing the movie (*Dimmock* v. *Secretary of State for Education and Skills,* 2007).

The principal source cited in Gore's movie and book, and arguably the reason it was well-received by much of the science community, was IPCC. There is no evidence IPCC ever complained about the misrepresentation of its report in the film or asked for corrections. Despite this and other documentation of the film's and book's many flaws (e.g., Lewis, 2007), Gore has never revised the book or even acknowledged the errors.

IPCC's reliability was crippled at birth, mandated by the UN Framework Convention on Climate Change (UNFCCC) to define climate change as human-caused climate change and to disregard naturally caused climate change. Since natural climate change is at the very center of the debate over whether human activity is influencing the climate and by how much, this essentially predetermined IPCC's conclusions. Tasked with finding a human impact on climate and calling on the nations of the world to do something about it, IPCC pursued its mission with fierce dedication.

IPCC's reports have been subjected to withering criticism by scientists and authors almost too numerous to count, including even high-profile editors and contributors to its reports (Seitz, 1996; Lindzen, 2012; Tol, 2014; Stavins, 2014) and no fewer than six rigorously researched books by one climate scientist, Patrick Michaels, former president of the American Association of State Climatologists, former program chair for the Committee on Applied Climatology of the American Meteorological Society, and a research professor of Environmental Sciences at University of Virginia for 30 years (Michaels, 1992, 2000, 2005a, 2005b, 2009, 2011). Michaels also was a contributing author and is a reviewer of IPCC's reports. Besides Michaels, see Singer, 1999; Essex and McKitrick, 2003; McIntyre and McKitrick, 2005; Pielke, R., 2010; Carter, 2010; Bell, 2011; and Vahrenholt and Lüning, 2015.

Others have pointed out IPCC's heavy reliance on environmental advocacy groups in the compilation of its official reports, using their personnel as lead authors and incorporating their publications – even newsletters – as source material (Laframboise, 2011). Scientists who participated in the latest IPCC report (AR5) described the process of producing the *Summary for Policymakers* as "exceptionally frustrating" and "one of the most extraordinary experiences of my academic life" (Economist, 2014).

Criticism hasn't come only from individual scientists. *Nature,* a prominent science journal, editorialized in 2013: "[I]t is time to rethink the IPCC. The organization deserves thanks and respect from all who care about the principle of evidence-based policy-making, but the current report should be its last mega-assessment." After describing the "exponential" growth of its reports and "truly breathtaking array of data" IPCC reports offer, the editors wrote, "Unfortunately, one thing that has not changed is that scientists cannot say with any certainty what rate of warming might be expected, or what effects humanity might want to prepare for, hedge against or avoid at all costs. In particular, the temperature range of the warming that would result from a doubling of atmospheric carbon dioxide levels is expected to be judged as 1.5– 4.5°C in next week's report – wider than in the last assessment and exactly what it was in the report of 1990. … Absent from next week's report, for instance, is recent and ongoing research on the rate of warming and what is – or is not – behind the plateau in average global temperatures that the world has experienced during the past 15 years. These questions have important policy implications, and the IPCC is the right body to answer them. But it need not wait six years to do so" (*Nature*, 2013).

In 2014, a science reporter for *Science,* published by the American Association for the Advancement of Science (AAAS), reported on political interference with IPCC's *Fifth Assessment Report*: "Although the underlying technical report from WGIII was accepted by the IPCC, final, heated negotiations among scientific authors and diplomats led to a substantial deletion of figures and text from the influential 'Summary for Policymakers' (SPM). … [S]ome fear that this redaction of content marks an overstepping of political interests, raising questions about division of labor between scientists and policy-makers and the need for new strategies in assessing complex science. Others argue that SPM should explicitly be coproduced with governments" (Wible, 2014). The subtitle of the article is "Did the 'Summary for Policymakers' become a summary by policy-makers?"

Later in 2014, after release of the Working Group III contribution to the Fifth Assessment Report, *Nature* reported critics "find the key conclusions unsurprising and short of detail. They say that the document sidesteps any hint of what specific countries, or groups of countries, should do to move towards clean energy systems. … Some researchers have long argued for a more pragmatic and diversified approach to climate change" (Schiermeier, 2014, p. 298).

Particularly harsh criticism of IPCC has come from the Amsterdam-based InterAcademy Council (IAC), which is made up of the presidents of many of the world's national science academies, the very academies defenders of IPCC often say endorse IPCC's findings. IAC conducted a thorough audit of IPCC in 2010 (IAC, 2010). Among its findings:

Fake confidence intervals: The IAC was highly critical of IPCC's method of assigning "confidence" levels to its forecasts, singling out "...the many statements in the Working Group II Summary for Policymakers that are assigned high confidence but are based on little evidence. Moreover, the apparent need to include statements of 'high confidence' (i.e., an 8 out of 10 chance of being correct) in the Summary for Policymakers led authors to make many vaguely defined statements that are difficult to refute, therefore making them of 'high confidence.' Such statements have little value" (p. 61).

Use of gray-sources: Too much reliance on unpublished and non-peer-reviewed sources (p. 63). Three sections of the IPCC's 2001 climate assessment cited peer-reviewed material only 36 percent, 59 percent, and 84 percent of the time.

Political interference: Line-by-line editing of the summaries for policymakers during "grueling Plenary session that lasts several days, usually culminating in an all-night meeting. Scientists and government representatives who responded to the Committee's questionnaire suggested changes to reduce opportunities for political interference with the scientific results..." (p. 64).

The use of secret data: "An unwillingness to share data with critics and enquirers and poor procedures to respond to freedom-of-information requests were the main problems uncovered in some of the controversies surrounding IPCC (Russell *et al.*, 2010; PBL, 2010). Poor access to data inhibits users' ability to check the quality of the data used and to verify the conclusions drawn..." (p. 68).

Selection of contributors is politicized: Politicians decide which scientists are allowed to participate in the writing and review process: "political considerations are given more weight than scientific qualifications" (p. 14).

Chapter authors exclude opposing views: "Equally important is combating confirmation bias—the tendency of authors to place too much weight on their own views relative to other views (Jonas *et al.*, 2001). As pointed out to the Committee by a presenter and some questionnaire respondents, alternative views are not always cited in a chapter if the Lead Authors do not agree with them..." (p. 18).

Need for independent review: "Although implementing the above recommendations would greatly strengthen the review process, it would not make the review process truly independent because the Working Group Co-chairs, who have overall responsibility for the preparation of the reports, are also responsible for selecting Review Editors. To be independent, the selection of Review Editors would have to be made by an individual or group not engaged in writing the report, and Review Editors would report directly to that individual or group (NRC, 1998, 2002)" (p. 21).

This is a damning critique. IPCC misrepresents its findings, does not properly peer review its reports, the selection of scientists who participate is politicized, the summary for policymakers is the product of late-night negotiations among governments and is not written by scientists, and more. The quotations above and the reference below are to a publicly circulated draft of IAC's final report, still available online (see reference). The final report was heavily edited to water down and perhaps hide the extent of problems uncovered by the investigators, itself evidence of still more misconduct. The report received virtually no press attention in the United States.

In 2012, IPCC issued a news release saying in part, "IPCC's 32nd session in Busan, Republic of Korea, in October 2010, adopted most of the IAC recommendations, and set up Task Groups to work on their implementation" (IPCC, 2012). One key recommendation, that a new Executive Committee be created that would include "three independent members," was almost comically disregarded: the committee was created, but all three slots were filled with IPCC employees (Laframboise, 2013). It is doubtful whether any other changes made at that time would have meaningfully affected the *Fifth Assessment Report*, which was already largely written. Media accounts of the release of AR5 once again told of late-night sessions with politicians and advocacy group representatives rewriting the *Summary for Policymakers*.

In conclusion, it is difficult to understand why IPCC reports still command the respect of anyone in the climate debate. They are political documents, not balanced or accurate summaries of the current state of climate science. They cannot provide reliable guidance to policymakers, economists, and climate scientists who put their trust in them.

References

Balling, R.C. 1992. *The Heated Debate: Greenhouse Predictions versus Climate Reality.* San Francisco, CA: Pacific Research Institute for Public Policy.

Bell, L. 2011. *Climate of Corruption: Politics and Power Behind the Global Warming Hoax.* Austin, TX: Greenleaf Book Group Press.

Bray, D. 2010. The scientific consensus of climate change revisited. *Environmental Science & Policy* **13:** 340– 350.

Bray, D. and Von Storch, H. 2010. A survey of climate scientists concerning climate science and climate change. http://www.academia.edu/2365610/A_Survey_of_Climate_Scientists_Concernin g_Climate_Science_and_Climate_Change.

Carlin, A. 2015. *Environmentalism Gone Mad: How a Sierra Club Activist and Senior EPA Analyst Discovered a Radical Green Energy Fantasy.* Mount Vernon, WA: Stairway Press.

Carter, R.M. 2010. *Climate: The Counter Consensus.* London, UK: Stacey International.

Darwall, R. 2013. *The Age of Global Warming: A History.* London, UK: Quartet Books Limited.

Dimmock vs. *Secretary of Education and Skills,* 2007. Case No. CO/3615/2007. Royal Courts of Justice, Strand, London. https://www.heartland.org/sites/default/files/justice_burton_decision.pdf.

Economist. 2014. Inside the sausage factory (May 10). http://www.economist.com/news/science-and-technology/21601813-scientists-v ersus-diplomats-intergovernmental-panel-climate

Essex, C. and McKitrick, R. 2003. *Taken By Storm: The Troubled Science,*

Policy and Politics of Global Warming. Toronto, Canada: Key Porter Books.

Gore, A. 2006. *An Inconvenient Truth*. Eamus, PA: Rodale Press.

Idso, S.B. 1982. *Carbon Dioxide: Friend or Foe?* Tempe, AZ: IBR Press.

Idso, S.B. 1989. *Carbon Dioxide and Global Change: Earth in Transition*. Tempe, AZ: IBR Press.

IAC, 2010. InterAcademy Council. Draft: *Climate Change Assessments: Review of the Processes & Procedures of IPCC*. Committee to Review the Intergovernmental Panel on Climate Change, October. The Hague, Netherlands. https://www.heartland.org/policy-documents/climate-change-assessments-revie w-processes-and-procedures-ipcc.

IPCC, 2012. IPCC completes review of processes and procedures. News Release (June 27). http://www.ipcc.ch/pdf/IAC_report/IAC_PR_Completion.pdf.

Jastrow, R., Nierenberg, W., and Seitz, F. 1990. *Scientific Perspectives on the Greenhouse Problem*. Ottawa, IL: The Marshall Press.

Jonas, E., Schulz-Hardt, S., Frey, D., and Thelen, N. 2001, Confirmation bias in sequential information search after preliminary decisions: An expansion of dissonance theoretical research on selective exposure to information. *Journal of Personality and Social Psychology* **80:** 557–571.

Laframboise, D. 2011. *The Delinquent Teenager Who was Mistaken for the World's Top Climate Expert*. Toronto, Canada: Ivy Avenue Press.

Laframboise, D. 2013. *Into the Dustbin: Rachendra Pachauri, the Climate Report & the Nobel Peace Prize*. CreateSpace Independent Publishing Platform.

Landsberg, H.E.1984. Global climatic trends. In: Simon, J.L. and Kahn, H. (Eds.) *The Resourceful Earth: A Response to 'Global 2000.'* New York, NY: Basil Blackwell Publisher Limited. 272–315.

Lewis, M. 2007. CEI Congressional Briefing Paper: Al Gore's Science Fiction. Washington, DC: Competitive Enterprise Institute.

Lindzen, R.S. 2012. Climate science: Is it currently designed to answer questions? *Euresis Journal* **2:** (Winter).

McIntyre, S. and McKitrick, R. 2005. Hockey sticks, principal components and spurious significance. *Geophysical Research Letters* **32:** L03710.

Michaels, P. 1992. *Sound and Fury: The Science and Politics of Global Warming.* Washington, DC: Cato Institute.

Michaels, P. 2000. *The Satanic Gases.* Washington, DC: Cato Institute.

Michaels, P. 2005a. *Meltdown: The Predictable Distortion of Global Warming by Scientists, Politicians, and the Media.* Washington, DC: Cato Institute.

Michaels, P. 2005b. *Shattered Consensus: The True State of Global Warming.* Lanham, MD: Rowman & Littlefield.

Michaels, P. 2009. *Climate of Extremes: Global Warming Science They Don't Want You to Know.* Washington, DC: Cato Institute.

Michaels, P. 2011. *Climate Coup: Global Warming's Invasion of Our Government and Our Lives.* Washington, DC: Cato Institute.

Moore, K., *et al.* 2014. *The Chain of Command: How a Club of Billionaires and Their Foundations Control the Environmental Movement and Obama's EPA.* Washington, DC: U.S. Senate Committee on Environment and Public Works.

Nature. 2013. The final assessment. Editorial. *Nature* **501:** (September 19), 281.

NRC. 1998. National Research Council. Peer Review in Environmental Technology Development Programs: The Department of Energy's Office of Science and Technology. Washington, DC: National Academy Press.

NRC. 2002. *Knowledge and Diplomacy: Science Advice in the United Nations System.* National Research Council. Washington, DC: National Academy Press.

PBL. 2010. *Assessing an IPCC Assessment: An Analysis of Statements on Projected Regional Impacts in the 2007 Report.* The Hague, Netherlands: Netherlands Environmental Assessment Agency.

Pielke Jr., R. 2010. *The Climate Fix: What Scientists and Politicians Won't Tell You About Global Warming.* New York, NY: Basic Books.

Pielke Jr., R., *et al.* 2009. Climate change: The need to consider human forcings besides greenhouse gases. *EOS* **90:** 45. 413–4.

Russell, M., Boulton, A.G., Clarke, P., Eyton, D., and Norton, J. 2010. *The Independent Climate Change E-mails Review Report to the University of East Anglia.* http://www.cce-review.org/pdf/final%20report.pdf. Last viewed on October 30, 2015.

Schiermeier, Q. 2014. IPCC report under fire. *Nature* **508:** (April 17) 298.

Seitz, F. 1996. A major deception on global warming. *The Wall Street Journal* (June 12).

Singer, S.F., Revelle, R., and Starr, C. 1992. What to do about greenhouse warming: Look before you leap. *Cosmos: A Journal of Emerging Issues* **5:** 2.

Singer, S.F. (Ed.) 1992. *The Greenhouse Debate Continued: An Analysis and Critique of IPCC Climate Assessment.* San Francisco, CA: ICS Press.

Singer, S.F. 1997. Rev. Ed. 1999. *Hot Talk Cold Science.* Oakland, CA: The Independent Institute.

Singer, S.F. 1989. (Ed.) *Global Climate Change.* New York, NY: Paragon House.

Solomon, L. 2008. *The Deniers: The World Renowned Scientists Who Stood Up Against Global Warming Hysteria, Political Persecution, and Fraud**And those who are too fearful to do so.* Washington, DC: Richard Vigilante Books.

Stavins, R. 2014. Is IPCC government approval process broken? Website (author's blog). http://www.robertstavinsblog.org/2014/04/25/is-the-ipcc-government-approval-process-broken-2/. Last viewed September 25, 2015.

Tol, R. 2014. IPCC again. Website (author's blog). http://richardtol.blogspot.nl/2014/04/ipcc-again.html. Last viewed September 25, 2015.

Vahrenholt, F. and Lüning, S. 2015. *The Neglected Sun: Why the Sun Precludes Climate Catastrophe.* Second English Edition. Chicago, IL: The Heartland Institute.

Weart, S. 2015. Climate change impacts: The growth of understanding. *Physics Today* (September 15). DOI: http://dx.doi.org/10.1063/PT.3.2914.

Wible, B. 2014. IPCC lessons from Berlin. *Science* **345:** 6192 (July 4) 34.

Bias

Climate scientists, like all humans, can be biased. Origins of bias include careerism, grant-seeking, political views, and confirmation bias.

Bias is another reason for disagreement among scientists and other writers on climate change. Scientists, no less than other human beings, bring their personal beliefs and interests to their work and sometimes make decisions based on them that direct their attention away from research findings that would contradict their opinions. Bias is often unconscious or overcome by professional ethics, but sometimes it leads to outright corruption.

Park *et al.* (2013), in a paper published in *Nature,* summarized research on publication bias, careerism, data fabrication, and fraud to explain how scientists converge on false conclusions. They write, "Here we show that even when scientists are motivated to promote the truth, their behaviour may be influenced, and even dominated, by information gleaned from their peers' behaviour, rather than by their personal dispositions. This phenomenon, known as herding, subjects the scientific community to an inherent risk of converging on an incorrect answer and raises the possibility that, under certain conditions, science may not be self-correcting."

Freedman (2010) identified a long list of reasons why experts are so often wrong, including pandering to audiences or clients, lack of oversight, reliance on flawed evidence provided by others, and failure to take into account important confounding variables.

John P.A. Ioannidis, professor of medicine and of health research and policy at Stanford University School of Medicine and a professor of statistics at Stanford University School of Humanities and Sciences, in a series of articles published in journals including the *Journal of the American Medical Association* (JAMA), revealed most published research in the health care field cannot be replicated or is likely to be contradicted by later publications (Ioannidis, 2005a, 2005b; Ioannidis and Trikalinos, 2005; Ioannidis, 2012). His most frequently cited work is titled "Why most published research findings are false."

Ioannidis's work generated widespread awareness that peer review is no guarantee of the accuracy or value of a research paper. In fact, he found that the likelihood of research being contradicted was highest with the most prestigious journals including *Nature, Science,* and *JAMA*. Springer, a major publisher of science journals, recently announced it was removing 16 papers it had published that were generated by a computer program called SCIgen that were simply gibberish (*Nature*, 2014). Much to their credit, these journals and academic institutions claim to be engaged in considerable soul-searching and efforts to reform a peer-review process that is plainly broken.

This controversy has particular relevance to the climate change debate due to "Climategate," the release of emails exchanged by prominent climate scientists discussing efforts to exclude global warming skeptics from journals, punish editors who allowed skeptics' articles to appear, stonewall requests for original data, manipulate data, and rush into publication articles refuting or attempting to discredit scientists who disagree with IPCC's findings (Montford, 2010; Sussman, 2010; Michaels, 2011, chapter 2). The scandal received little press attention in the United States. Journals such as *Nature* take the scandal over peer-review corruption seriously when it involves other topics (Ferguson *et al.*, 2014), are curiously silent about its occurrence in the climate change literature.

Scientists, especially those in charge of large research projects and laboratories, have a financial incentive to seek more funding for their programs. They are not immune to having tunnel vision regarding the importance of their work and employment. Each believes his or her mission is more significant and essential relative to other budget priorities.

To obtain funding (and more funding), it helps scientists immensely to have the public – and thus Congress and potentially private funders – worried about the critical nature of the problems they study. This incentive makes it less likely researchers will interpret existing knowledge or present their findings in a way that reduces public concern (Lichter and Rothman, 1999; Kellow, 2007; Kabat, 2008). As a result, scientists often gravitate toward emphasizing worst-case scenarios, though there may be ample evidence to the contrary. This bias of alarmism knows no political bounds, affecting both liberal Democrats and conservative Republicans (Berezow and Campbell, 2012; Lindzen, 2012).

Alarmists in the climate debate seem to recognize only one possible source of bias, and that is funding from "the fossil fuel industry." The accusation permeates any conversation of the subject, perhaps second only to the "consensus" claim, and the two are often paired, as in "only scientists paid by the fossil fuel industry dispute the overwhelming scientific consensus." The accusation doesn't work for many reasons:

■ There has never been any evidence of a climate scientist accepting money from industry to take a position or change his or her position in the climate debate (Cook, 2014);

■ Vanishingly few global warming skeptics have ever been paid by the fossil fuel industry. Certainly not more than a tiny fraction of the 31,478 American scientists who signed the Global Warming Petition or the thousands of meteorologists and climate scientists reported in Chapter 1 who tell survey-takers they do not agree with IPCC;

■ Funding of alarmists by government agencies, liberal foundations, environmental advocacy groups, and the alternative energy industry exceeds funding from the fossil fuel industry by two, three, or even four orders of magnitude (Butos and McQuade, 2015). Does government and interest-group funding of alarmists not also have a "corrupting" influence on its recipients?

■ The most prominent organizations supporting global warming skepticism get little if any money from the fossil fuel industry. Their support comes overwhelmingly from individuals (and their foundations) motivated by concern over the apparent corruption of science taking place and the enormous costs it is imposing on the public.

In the text of her speech to the British House of Lords cited earlier, climate scientist Judith Curry wrote, "I am very concerned that climate science is becoming biased owing to biases in federal funding priorities and the institutionalization by professional societies of a particular ideology related to climate change. Many scientists, and institutions that support science, are becoming advocates for UN climate policies, which is leading scientists into overconfidence in their assessments and public statements and into failures to respond to genuine criticisms of the scientific consensus. In short, the climate science establishment has become intolerant to disagreement and debate, and is attempting to marginalize and de-legitimized dissent as corrupt or ignorant" (Curry, 2015).

Money probably isn't what motivates Mike Hulme, now professor of climate and culture in the Department of Geography at King's College London. He was formerly professor of climate change in the School of Environmental Sciences at the University of East Anglia, a contributor to IPCC reports, and author of *Why We Disagree About Climate Change* (Hulme, 2009). Hulme was cited earlier in this chapter admitting to great uncertainties in climate science, yet he eagerly endorses and promotes IPCC's claims. Why does he do that?

In his book, Hulme calls climate change "a classic example of ... 'post-normal science,'" which he defines (quoting Silvio Funtowicz and Jerry Ravetz) as "the application of science to public issues where 'facts are uncertain, values in dispute, stakes high and decisions urgent.'" Issues that fall into this category, he says, are no longer subject to the cardinal requirements of true science: skepticism, universalism, communalism, and disinterestedness. Instead of experimentation and open debate, post-normal science says "consensus" brought about by deliberation among experts determines what is true, or at least true enough for the time being to direct public policy decisions.

The merits and demerits of post-normal science can be debated, but it undoubtedly has one consequence of significance in the climate change debate: scientists are no longer responsible for actually doing science themselves, such as testing hypotheses, studying data, and confronting data or theories that contradict the "consensus" position. Scientists simply "sign onto" IPCC's latest report and are free to indulge their political biases. Hulme is quite open about his. He wrote, "The idea of climate change should be seen as an intellectual resource around which our collective and personal identities and projects can form and take shape. We need to ask not what we can do for climate change, but to ask what climate change can do for us" (p. 326).

In his book, Hulme says "because the idea of climate change is so plastic, it can be deployed across many of our human projects and can serve many of our psychological, ethical, and spiritual needs." Hulme describes himself as a social-democrat so his needs include sustainable development, income redistribution, population control, and social justice. By focusing on these "needs," how can Hulme objectively evaluate the anthropogenic global warming hypothesis?

Like the late Stephen Schneider, who once said "to reduce the risk of potentially disastrous climate change … we need to get some broad based support, to capture the public's imagination. That, of course, means getting loads of media coverage. So we have to offer up scary scenarios, make simplified, dramatic statements, and make little mention of any doubts we might have" (Schneider, 1989), Hulme wrote, "We will continue to create and tell new stories about climate change and mobilise them in support of our projects." He suggests his fellow global warming alarmists promote four "myths," which he labels Lamenting Eden, Presaging Apocalypse, Constructing Babel, and Celebrating Jubilee.

This is unusual behavior for a scientist and disturbing for one working at high levels in IPCC. When Hulme talks about climate science, is he telling us the truth or one of his "myths"?

* * *

While it would be ideal if scientists could be relied upon to deliver the unvarnished truth about complex scientific matters to governments and voters, the truth is they almost always fall short. Ignorance of research outside their area of specialization, reliance on flawed authorities, bias, and outright corruption all contribute to unwarranted alarmism in the climate change debate.

References

Berezow, A.B. and Campbell, H. 2012. *Science Left Behind: Feel-Good Fallacies and the Rise of the Anti-Scientific Left.* Philadelphia, PA: PublicAffairs.

Butos, W.N. and McQuade, T.J. 2015. Causes and consequences of the climate science boom. *The Independent Review* **20:** 2 (Fall), 165–196.

Cook, R. 2014. Merchants of smear. *Heartland Policy Brief* (September). Chicago, IL: The Heartland Institute.

Curry, J. 2015. State of the climate debate in the U.S. Remarks to the U.K. House of Lords, June 15. Climate Etc. Website. http://judithcurry.com/2015/06/15/state-of-the-climate-debate-in-the-u-s/. Last viewed on October 1, 2015.

Ferguson, C., Marcus, A., and Oransky, I. 2014. Publishing: The peer-review scam. *Nature* **515:** 480–482. doi:10.1038/515480a.

Freedman, D.H. 2010. *Wrong: Why Experts Keep Failing Us – And How to Know When Not to Trust Them.* New York, NY: Little, Brown and Company.

Hulme, M. 2009. *Why We Disagree About Climate Change: Understanding Controversy, Inaction and Opportunity.* New York, NY: Cambridge University Press.

Ioannidis, J.P.A. 2005a. Contradicted and initially stronger effects in highly cited clinical research. *Journal of the American Medical Association* **294:** 218–228.

Ioannidis, J.P.A. 2005b. Why most published research findings are false. *PLOS Medicine* **2:** e124.

Ioannidis, J.P.A. 2012. Scientific inbreeding and same-team replication: Type D personality as an example. *Journal of Psychosomatic Research* **73:** 408–410.

Ioannidis, J.P. and Trikalinos, T.A. 2005. Early extreme contradictory estimates may appear in published research: the Proteus phenomenon in molecular genetics research and randomized trials. *Journal of Clinical Epidemiology* **58.**

Kabat, G.C. 2008. *Hyping Health Risks: Environmental Hazards in Daily Life and the Science of Epidemiology.* New York, NY: Columbia University Press.

Kellow, A. 2007. *Science and Public Policy: The Virtuous Corruption of Virtual Environmental Science.* Northampton, MA: Edward Elgar Publishing.

Lichter, S.R. and Rothman, S. 1999. *Environmental Cancer – A Political Disease?* New Haven, CT: Yale University Press.

Lindzen, R.S. 2012. Climate science: Is it currently designed to answer questions? *Euresis Journal* **2:** (Winter).

Michaels, P. 2011. (Ed.) *Climate Coup: Global Warming's Invasion of Our government and Our Lives.* Washington, DC: Cato Institute.

Montford, A.W. 2010. *Hockey Stick Illusion: Climategate and the Corruption of*

Science. London, UK: Stacey International.

Nature. 2014. Gibberish papers. [news item] *Nature* **507:** 6. 13.

Park, I-U., Peacey, M.W., and Munafo, M.R. 2014. Modelling the effects of subjective and objective decision making in scientific peer review. *Nature* **506:** 6 (February). 93–96.

Schneider, S. 1989. Interview. *Discover* (October). 45–48.

Sussman, B. 2010. *Climategate: A Veteran Meteorologist Exposes the Global Warming Scam.* Washington, DC: Worldnet Daily.

3

Scientific Method vs. Political Science

Key findings of this section include the following:

- The hypothesis implicit in all IPCC writings, though rarely explicitly stated, is that dangerous global warming is resulting, or will result, from human-related greenhouse gas emissions.

- The null hypothesis is that currently observed changes in global climate indices and the physical environment, as well as current changes in animal and plant characteristics, are the result of natural variability.

- In contradiction of the scientific method, IPCC assumes its implicit hypothesis is correct and that its only duty is to collect evidence and make plausible arguments in the hypothesis's favor.

The Missing Null Hypothesis

Although IPCC's reports are voluminous and their arguments impressively persistent, it is legitimate to ask whether that makes them good science. In order to conduct an investigation, scientists must first formulate a falsifiable hypothesis to test. The hypothesis implicit in all IPCC writings, though rarely explicitly stated, is that dangerous global warming is resulting, or will result, from human-related greenhouse gas emissions.

In considering any such hypothesis, an alternative and null hypothesis must be entertained, which is the simplest hypothesis consistent with the known facts. Regarding global warming, the null hypothesis is that currently observed changes in global climate indices and the physical environment are the result of natural variability. To invalidate this null hypothesis requires, at a minimum, direct evidence of human causation of specified changes that lie outside usual, natural variability. Unless and until such evidence is adduced, the null hypothesis is assumed to be correct.

Science does not advance by consensus, a show of hands, or even persuasion. It advances by individual scientists proposing testable hypotheses, examining data to see if they disprove a hypothesis, and making those data available to other unbiased researchers to see if they arrive at similar conclusions. Disagreement is the rule and consensus is the exception in most academic disciplines. This is because science is a process leading to ever-greater certainty, necessarily implying that what is accepted as true today will likely not be accepted as true tomorrow. Albert Einstein was absolutely right when he said, "No amount of experimentation can ever prove me right; a single experiment can prove me wrong" (Einstein, 1996).

In contradiction of the scientific method, IPCC assumes its implicit hypothesis is correct and that its only duty is to collect evidence and make plausible arguments in the hypothesis's favor. One probable reason for this behavior is that the United Nations protocol under which IPCC operates defines climate change as "a change of climate which is attributed directly or indirectly to human activity that alters the composition of the global atmosphere and which is in addition to natural climate variability observed over comparable time periods" (United Nations, 1994, Article 1.2). Not surprisingly, directing attention to only the effects of human greenhouse gas emissions has resulted in IPCC failing to provide a thorough analysis of climate change.

References

Einstein, A. 1996. Quoted in A. Calaprice, *The Quotable Einstein.* Princeton, MA: Princeton University Press. p. 224.

United Nations. 1994. Framework convention on climate change. http://unfccc.int/resource/docs/convkp/conveng.pdf.

Models, Postulates, and Circumstantial Evidence

IPCC offers three lines of reasoning in defense of its hypothesis: global climate model projections, a series of postulates or assumptions, and appeals to circumstantial evidence. The specific arguments are summarized in Figure 2.

Figure 2
IPCC's Three Lines of Argument

Global Climate Model Projections

IPCC modelers assume Global Climate Models (GCMs) are based on a perfect knowledge of all climate forcings and feedbacks. They then assert:

- A doubling of atmospheric CO_2 would cause warming of up to 6°C.
- Human-related CO_2 emissions caused an atmospheric warming of at least 0.3°C over the past 15 years.
- Enhanced warming (a "hot spot") should exist in the upper troposphere in tropical regions.
- Both poles should have warmed faster than the rest of Earth during the late twentieth century.

Postulates

Postulates are statements that assume the truth of an underlying fact that has not been independently confirmed or proven. IPCC postulates:

- The warming of the twentieth century cannot be explained by natural variability.
- The late twentieth century warm peak was of greater magnitude than previous natural peaks.
- Increases in atmospheric CO_2 precede, and then force, parallel increases in temperature.
- Solar forcings are too small to explain twentieth century warming.
- A future warming of 2°C or more would be net harmful to the biosphere and human well-being.

Circumstantial Evidence

Circumstantial evidence does not bear directly on the matter in dispute but refers to circumstances from which the occurrence of the fact might be inferred. IPCC cites the following circumstantial evidence it says is consistent with its hypothesis:

- Unusual melting is occurring in mountain glaciers, Arctic sea ice, and polar icecaps.
- Global sea level is rising at an enhanced rate and swamping tropical coral atolls.
- Droughts, floods, and monsoon variability and intensity are increasing.
- Global warming is leading to more, or more intense, wildfires, rainfall, storms, hurricanes, and other extreme weather events.
- Unusual melting of Boreal permafrost or sub-seabed gas hydrates is causing warming due to methane release.

Source: Summary for Policymakers, *Climate Change Reconsidered II: Physical Science* (Chicago, IL: The Heartland Institute, 2013).

All three lines of reasoning depart from proper scientific methodology. Global climate models produce meaningful results only if we assume we already know perfectly how the global climate works, and most climate scientists say we do not (Bray and von Storch, 2010; Strengers, Verheggen, and Vringer, 2015). Moreover, it is widely recognized that climate models are not designed to produce predictions of future climate but rather what-if projections of many alternative possible futures (Trenberth, 2009).

Postulates, commonly defined as "something suggested or assumed as true as the basis for reasoning, discussion, or belief," can stimulate relevant observations or experiments but more often are merely assertions that are difficult or impossible to test (Kahneman, 2011). IPCC expresses "great confidence" and even "extreme confidence" in its assumptions, but it cannot apply a statistical confidence level because they are statements of opinion and not of fact. This is not the scientific method.

Circumstantial evidence, or observations, in science are useful primarily to falsify hypotheses and cannot prove one is correct (Popper, 1965, p. vii). It is relatively easy to assemble reams of "evidence" in favor of a point of view or opinion while ignoring inconvenient facts that would contradict it, a phenomenon called "confirmation bias." The only way to avoid confirmation bias is independent review of a scientist's work by other scientists who do not have a professional, reputational, or financial stake in whether the hypothesis is confirmed or disproven. As documented in Chapter 2, this sort of review is conspicuously absent in the climate change debate. Those who attempt to exercise it find themselves demonized, their work summarily rejected by academic journals, and worse.

Facing such criticism of its methodology and a lack of compelling evidence of dangerous warming, IPCC's defenders often invoke the precautionary principle. The principle states: "Where there are threats of serious or irreversible damage, lack of full scientific certainty shall not be used as a reason for postponing cost-effective measures to prevent environmental degradation" (United Nations, 1992, Principle 15). This is a sociological precept rather than a scientific one and lacks the intellectual rigor necessary for use in policy formulation (Goklany, 2001).

The hypothesis of human-caused global warming comes up short not merely of "full scientific certainty" but of reasonable certainty or even plausibility. The weight of evidence now leans heavily against the theory. Invoking the precautionary principle does not lower the required threshold for evidence to be regarded as valid nor does it answer the most important questions about the causes and consequences of climate change. Scientific principles acknowledge the supremacy of experiment and observation and do not bow to instinctive feelings of alarm or claims of a supposed scientific "consensus" (Legates et al., 2013). The formulation of effective public environmental policy must be rooted in evidence-based science, not an over-abundance of precaution (More and Vita-More, 2013; U.K. House of Commons Science and Technology Committee, 2006).

Contradictions about methodology and the verity of claimed facts make it difficult for unprejudiced lay persons to judge for themselves where the truth actually lies in the global warming debate. This is one of the primary reasons why politicians and commentators rely so heavily on supposedly authoritative statements issued by one side or another in the public discussion. Arguing from authority, however, is the antithesis of the scientific method. Attempting to stifle debate by appealing to authority hinders rather than helps scientific progress and understanding.

References

Bray, D. and von Storch, H. 2010. CliSci2008: A survey of the perspectives of climate scientists concerning climate science and climate change. GKSS-Forschungszentrum Geesthacht GmbH. http://ncseprojects.org/files/pub/polls/2010--Perspectives_of_Climate_Scientists _Concerning_Climate_Science_&_Climate_Change_.pdf.

Goklany, I.M. 2001. *The Precautionary Principle: A Critical Appraisal of Environmental Risk Assessment.* Washington, DC: Cato Institute.

Kahneman, D. 2011. *Thinking, Fast and Slow.* New York, NY: Macmillan.

Legates, D.R., Soon, W., Briggs, W.M., and Monckton, C. 2013. Climate consensus and 'misinformation': A rejoinder to 'Agnotology, scientific consensus, and the teaching and learning of climate change.' *Science & Education,* doi 10.1007/s1119-013-9647-9.

More, M. and Vita-More, N. 2013. (Eds.) *The Transhumanist Reader: Classical and Contemporary Essays on the Science, Technology, and Philosophy of the Human Future.* New York, NY: John Wiley & Sons, Inc.

Popper, K. 1965. *Conjectures and Refutations: The Growth of Scientific Knowledge.* Second edition. New York, NY: Harper and Row, Publishers.

Strengers, B., Verheggen, B., and Vringer, K. 2015. Climate science survey questions and responses. April 10. PBL Netherlands Environmental Assessment Agency. http://www.pbl.nl/sites/default/files/cms/publicaties/pbl-2015-climate-science-survey-questions-and-responses_01731.pdf.

Trenberth, K.E. 2009. Climate feedback: predictions of climate. *Nature.* Blog. April 11.

U.K. House of Commons Science and Technology Committee. 2006. Scientific Advice, Risk and Evidence Based Policy Making. Seventh Report of Session 2005–06. http://www.publications.parliament.uk/pa/cm200506/cmselect/cmsctech/900/900 -i.pdf.

United Nations. 1992. Report of the United Nations conference on environmental development (Rio de Janeiro, June 3–14, 1992). http://www.un.org/documents/ga/conf151/aconf15126-1annex1.htm.

4

—

Flawed Projections

Key findings in this section include the following:

- The United Nations' Intergovernmental Panel on Climate Change (IPCC) and virtually all the governments of the world depend on global climate models (GCMs) to forecast the effects of human-related greenhouse gas emissions on the climate.

- GCMs systematically over-estimate the sensitivity of climate to carbon dioxide (CO_2), many known forcings and feedbacks are poorly modeled, and modelers exclude forcings and feedbacks that run counter to their mission to find a human influence on climate.

- The Nongovernmental International Panel on Climate Change (NIPCC) estimates a doubling of CO_2 from pre-industrial levels (from 280 to 560 ppm) would likely produce a temperature forcing of 3.7 Wm^{-2} in the lower atmosphere, for about ~1°C of *prima facie* warming.

- Four specific forecasts made by GCMs have been falsified by real-world data from a wide variety of sources. In particular, there has been no global warming for some 18 years.

Why Computer Models Are Flawed

In contrast to the scientific method, IPCC and virtually all national governments in the world rely on computer models, called global climate models or GCMs, to represent speculative thought experiments by modellers who often lack a detailed understanding of underlying processes. The results of GCMs are only as reliable as the data and theories "fed" into them, which scientists widely recognize as being seriously deficient. If natural climate forcings and feedback are not perfectly understood, then GCMs become little more than an exercise in curve-fitting, or changing parameters until the outcomes match the modeller's expectations. As John von Neumann is reported to have once said, "with four parameters I can fit an elephant, and with five I can make him wiggle his trunk" (Dyson, 2004).

The science literature is replete with admissions by leading climate modellers that forcings and feedback are not sufficiently well understood, that data are insufficient or too unreliable, and that computer power is insufficient to resolve important climate processes. Many important elements of the climate system, including atmospheric pressure, wind, clouds, temperature, precipitation, ocean currents, sea ice, and permafrost, cannot be properly simulated by the current generation of models.

The major known deficiencies include model calibration, non-linear model behavior, and the omission of important natural climate-related variability. Model calibration is faulty as it assumes all temperature rise since the start of the industrial revolution has resulted from human CO_2 emissions. In reality, major human-related emissions commenced only in the mid-twentieth century.

More facts about climate models and their limitations reported in Chapter 1 of *Climate Change Reconsidered-II: Physical Science* are reported in Figure 3.

Figure 3
Key Facts about Global Climate Models

- Climate models generally assume a climate sensitivity of 3°C for a doubling of CO_2 above preindustrial values, whereas meteorological observations are consistent with a sensitivity of 1°C or less.

- Climate models underestimate surface evaporation caused by increased temperature by a factor of 3, resulting in a consequential under-estimation of global precipitation.

■ Climate models inadequately represent aerosol-induced changes in infrared (IR) radiation, despite studies showing different mineral aerosols (for equal loadings) can cause differences in surface IR flux between 7 and 25 Wm^{-2}.

■ Deterministic climate models have inherent properties that make dynamic predictability impossible; introduction of techniques to deal with this (notably parameterization) introduces bias into model projections.

■ Limitations in computing power restrict climate models from resolving important climate processes; low-resolution models fail to capture many important regional and lesser-scale phenomena such as clouds.

■ Model calibration is faulty, as it assumes all temperature rise since the start of the industrial revolution has resulted from human CO_2 emissions; in reality, major human-related emissions commenced only in the mid-twentieth century.

■ Non-linear climate models exhibit chaotic behavior. As a result, individual simulations ("runs") may show differing trend values.

■ Internal climate oscillations (AMO, PDO, etc.) are major features of the historic temperature record; climate models do not even attempt to simulate them.

■ Climate models fail to incorporate the effects of variations in solar magnetic field or in the flux of cosmic rays, both of which are known to significantly affect climate.

Source: "Chapter 1. Global Climate Models and Their Limitations," *Climate Change Reconsidered II: Physical Science* (Chicago, IL: The Heartland Institute, 2013).

Forcings and Feedbacks

The discussion in the previous section of why global climate models are flawed included references to some of the forcings and feedbacks that are poorly modeled and likely to make models unreliable. In many of these cases, climate scientists are substituting opinions or best guesses for data. As serious as that problem is, it is made worse by the exclusion of forcings and feedbacks that are well documented in the scientific literature. Many of these run counter to the goal of many modelers to find a human influence on climate and so are ignored.

Among the forcings and feedbacks IPCC has failed to take into account are increases in low-level clouds in response to enhanced atmospheric water vapor, ocean emissions of dimethyl sulfide (DMS), and the presence and total cooling effect of both natural and industrial aerosols. These processes and others are likely to offset most or even all of any warming caused by rising CO_2 concentrations. Figure 4 summarizes these and other findings about forcings and feedbacks appearing in Chapter 2 of *Climate Change Reconsidered-II: Physical Science.*

Figure 4
Key Facts about Temperature Forcings and Feedbacks

- A doubling of CO_2 from pre-industrial levels (from 280 to 560 ppm) would likely produce a temperature forcing of 3.7 Wm^{-2} in the lower atmosphere, for about ~1°C of *prima facie* warming.

- IPCC models stress the importance of positive feedback from increasing water vapor and thereby project warming of ~3–6°C, whereas empirical data indicate an order of magnitude less warming of ~0.3–1.0°C.

- In ice core samples, changes in temperature precede parallel changes in atmospheric CO_2 by several hundred years; also, temperature and CO_2 are uncoupled through lengthy portions of the historical and geological records; therefore CO_2 cannot be the primary forcing agent for most temperature changes.

- Atmospheric methane (CH_4) levels for the past two decades fall well below the values projected by IPCC in its assessment reports. IPCC's temperature projections incorporate these inflated CH_4 estimates and need downward revision accordingly.

- The thawing of permafrost or submarine gas hydrates is not likely to emit dangerous amounts of methane at current rates of warming.

- Nitrous oxide (N_2O) emissions are expected to fall as CO_2 concentrations and temperatures rise, indicating it acts as a negative climate feedback.

- Other negative feedbacks on climate sensitivity that are either discounted or underestimated by IPCC include increases in low-level clouds in response to enhanced atmospheric water vapor, increases in ocean emissions of dimethyl sulfide (DMS), and the presence and total cooling effect of both natural and industrial aerosols.

Source: "Chapter 2. Forcings and Feedbacks," *Climate Change Reconsidered II: Physical Science* (Chicago, IL: The Heartland Institute, 2013).

Yet another deficiency in GCMs is that non-linear climate models exhibit chaotic behavior. As a result, individual simulations ("runs") may show differing trend values (Singer, 2013b). Internal climate oscillations (Atlantic Multidecadal Oscillation (AMO), Pacific Decadal Oscillation (PDO), etc.) are major features of the historic temperature record, yet GCMs do not even attempt to simulate them. Similarly, the models fail to incorporate the effects of variations in solar magnetic field or in the flux of cosmic rays, both phenomena known to significantly affect climate.

We conclude the current generation of GCMs is unable to make accurate projections of climate even 10 years ahead, let alone the 100-year period that has been adopted by policy planners. The output of such models should therefore not be used to guide public policy formulation until they have been validated and shown to have predictive value.

Failed Forecasts

Four specific forecasts made by GCMs have been falsified by real-world data from a wide variety of sources:

Failed Forecast #1: A doubling of atmospheric CO_2 would cause warming between 3°C and 6°C.

The increase in radiative forcing produced by a doubling of atmospheric CO_2 is generally agreed to be 3.7 Wm^{-2}. Equating this forcing to temperature requires taking account of both positive and negative feedbacks. IPCC models incorporate a strong positive feedback from increasing water vapor but exclude negative feedbacks such as a concomitant increase in low-level clouds – hence they project a warming effect of 3°C or more.

IPCC ignores mounting evidence that climate sensitivity to CO_2 is much lower than its models assume (Spencer and Braswell, 2008; Lindzen and Choi, 2011). Monkton et al. cited 27 peer-reviewed articles "that report climate sensitivity to be below current central estimates" (Monckton et al., 2015). Their list of sources appears in Figure 5.

Figure 5
Research Finding Climate Sensitivity Is Less than Assumed by IPCC

Michaels, P.J., Knappenberger, P.C., Frauenfeld, O.W. *et al.* 2002. Revised 21st century temperature projections. *Climate Research* **23:** 1–9.

Douglass, D.H., Pearson, B.D., and Singer, S.F. 2004. Altitude dependence of atmospheric temperature trends: climate models versus observation. *Geophysical Research Letters* **31:** L13208, doi:10.1029/2004GL020103.

Landscheidt, T. 2003. New Little Ice Age instead of global warming? *Energy & Environment* **14:** 2, 327–350.

Chylek, P. and Lohmann, U. 2008. Aerosol radiative forcing and climate sensitivity deduced from the Last Glacial Maximum to Holocene transition. *Geophysical Research Letters* **35:** L04804, doi:10.1029/2007GL032759.

Monckton of Brenchley, C. 2008. Climate sensitivity reconsidered. *Physics & Society* **37:** 6–19.

Douglass, D.H. and Christy, J.R. 2009. Limits on CO_2 climate forcing from recent temperature data of earth. *Energy & Environment* **20:** 1–2.

Lindzen, R.S. and Choi, Y-S. 2009. On the determination of climate feedbacks from ERBE data. *Geophysical Research Letters* **36:** L16705, doi:10.1029/2009GL039628.

Spencer, R.W. and Braswell, W.D. 2010. On the diagnosis of radiative feedback in the presence of unknown radiative forcing. *Journal of Geophysical Research* **115:** D16109, doi:10.1029/2009JD013371.

Annan, J.D. and Hargreaves, J.C. 2011. On the generation and interpretation of probabilistic estimates of climate sensitivity. *Climate Change* **104:** 324–436.

Lindzen, R.S. and Choi, Y-S. 2011 On the observational determination of climate sensitivity and its implications. *Asia-Pacific Journal of Atmospheric Sciences* **47:** 377-390.

Monckton of Brenchley, C. 2011. Global brightening and climate sensitivity. In: Zichichi, A. and Ragaini, R. (Eds.) *Proceedings of the 45th Annual International Seminar on Nuclear War and Planetary Emergencies*, World Federation of Scientists. London, UK: World Scientific.

Schmittner, A., Urban, N.M., Shakun, J.D., *et al.* 2011. Climate sensitivity estimated from temperature reconstructions of the last glacial maximum. *Science* **334:** 1385–1388, doi:10.1126/science.1203513.

Spencer, R.W. and Braswell, W.D. 2011. On the misdiagnosis of surface temperature feedbacks from variations in Earth's radiant-energy balance. *Remote Sensing* **3:** 1603-1613, doi:10.3390/rs3081603.

Aldrin, M., Holden, M., Guttorp, P., *et al.* 2012. Bayesian estimation of climate sensitivity based on a simple climate model fitted to observations of hemispheric temperature and global ocean heat content. *Environmetrics* **23:** 253-271, doi: 10.1002/env.2140.

Hargreaves, J.C., Annan, J.D., Yoshimori, M., *et al.* 2012. Can the last glacial maximum constrain climate sensitivity? *Geophysical Research Letters* **39:** L24702, doi:10.1029/2012GL053872.

Ring, M.J., Lindner, D., Cross, E.F., *et al.* 2012. Causes of the global warming observed since the 19th century. *Atmospheric and Climate Sciences* **2:** 401–415, doi: 10.4236/acs.2012.24035.

van Hateren, J.H. 2012. A fractal climate response function can simulate global average temperature trends of the modern era and the past millennium. *Climate Dynamics* **40:** 2651–2670, doi:10.1007/s00382-012-1375-3.

Lewis, N. 2013. An objective Bayesian improved approach for applying optimal fingerprint techniques to estimate climate sensitivity. *Journal of Climate* **26:** 7414–7429, doi:10.1175/JCLI-D-12-00473.1.

Masters, T. 2013. Observational estimates of climate sensitivity from changes in the rate of ocean heat uptake and comparison to CMIP5 models. *Climate Dynamics* **42:** 2173–2181, doi:101007/s00382-013-1770-4.

Otto, A., Otto, F.E.L., Boucher, O., *et al.* 2013. Energy budget constraints on climate response. *Nature Geoscience* **6:** 415-416, diuL19,1938/ngeo1836.

Spencer, R.W. and Braswell, W.D. 2013. The role of ENSO in global ocean temperature changes during 1955-2011 simulated with a 1D climate model. *Asia-Pacific Journal of Atmospheric Sciences* **50:** 229-237, doi:10.1007/s13143-014-0011-z.

Lewis, N. and Curry, J.A. 2014. The implications for climate sensitivity of AR5 forcing and heat uptake estimates. *Climate Dynamics* **10:** 1007/s00382-014-2342-y.

Loehle, C. 2014. A minimal model for estimating climate sensitivity. *Ecological Modelling* **276:** 80–84, doi:10.1016/j.ecolmodel.2014.01.006.

McKitrick, R. 2014. HAC-robust measurement of the duration of a trendless subsample in a global climate time series. *Open Journal of Statistics* **4:** 527-535, doi:10.4236/ojs.2014.47050.

Monckton of Brenchley, C. 2014. Political science: drawbacks of apriorism in intergovernmental climatology. *Energy & Environment* **25:** 1177–1204.

Skeie, R.B., Berntsen, T., Aldrin, M., *et al.* 2015. A lower and more constrained estimate of climate sensitivity using updated observations and detailed radiative forcing time series. *Earth System Dynamics* **5:** 139–175, doi:10.5194/esd-5-139-2014.

Lewis, N. 2015. Implications of recent multimodel attribution studies for climate sensitivity. *Climate Dynamics* doi:10.1007/s00382-015-2653-7RSS.

Source: Monckton, C., Soon, W. W-H., Legates, D.R., and Briggs, W.M. 2015. Keeping it simple: the value of an irreducibly simple climate model. *Science Bulletin* **60:** 15, 1378–1390, footnotes 7 to 33.

Failed Forecast #2: CO_2 caused an atmospheric warming of at least 0.3°C over the past 15 years.

The global climate models relied on by IPCC predicted an atmospheric warming of at least 0.3°C during the first 15 years of the twenty-first century, but temperatures did not rise at all during that period. Figure 6 shows global temperatures from 1997 to 2015, based on satellite data compiled and reported by Remote Sensing Systems and interpreted by Monckton *et al.* (2015). They show a trend of -0.01°C from January 1997 to June 2015. Figure 7 vividly portrays the failure of GCMs to hindcast this trend.

Figure 6
RSS Monthly Global Mean Lower-troposphere Temperature Anomalies, January 1997 to June 2015

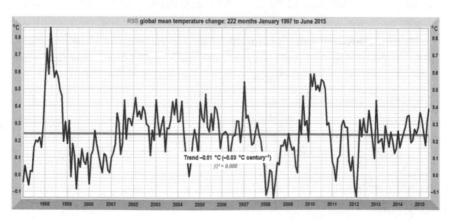

Source: Monckton *et al.*, 2015. Data from Mears and Wentz, 2009.

Figure 7. Linear Trends on Tropical Mid-troposphere Temperature Anomalies Projected by 73 Models and Measured by Two Coincident Observational Datasets, 1979–2012

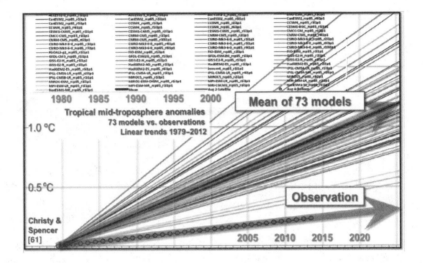

Source: Monckton *et al.*, 2015.

The absence of a warming trend for more than 15 years invalidates GCMs based on IPCC's assumptions regarding climate sensitivity to carbon dioxide. In its 2008 *State of the Climate* report, the National Oceanic and Atmospheric Administration (NOAA) reported, "Near zero and even negative trends are common for intervals of a decade or less in the simulations, due to the models internal climate variability. The simulations rule out (at the 95% level) zero trends for intervals of 15 yr or more, suggesting that an observed absence of warming of this duration is needed to create a discrepancy with the expected present-day warming rate" (Knight *et al.*, 2009). This "discrepancy" now exists, indeed now extends to 18 years without warming, and the models have been invalidated.

IPCC's authors compare the output of unforced (and incomplete) models with a dataset that represents twentieth century global temperature (HadCRUT, British Meteorological Office). Finding a greater warming trend in the dataset than in model projections, the false conclusion is then drawn that this "excess" warming must be caused by human-related greenhouse forcing. In reality, no excess warming has been demonstrated, first because this line of argument assumes models have perfect knowledge, information, and power, which they do not, and second, because a wide variety of datasets other than the HadCRUT global air temperature curve favored by IPCC do not exhibit a warming trend during the second half of the twentieth century. See Figure 8.

Figure 8
Lack of Evidence for Rising Temperatures

The difference in surface temperatures between 1942–1995 and 1979–97, as registered by datasets that represent land, oceanic, and atmospheric locations.

LAND SURFACE	Global (IPCC, HadCRUT)	+0.5° C
	United States (GISS)	~zero
OCEAN	Sea surface temperature (SST)[1]	~zero
	SST Hadley NMAT	~zero
ATMOSPHERE	Satellite MSU (1979–1997)	~zero
	Hadley radiosondes (1979–97)	~zero
PROXIES	Mostly land surface temperature[2]	~zero

Unless otherwise indicated, data are drawn from the nominated government agencies.

Sources: [1] Gouretski et al., 2012; [2] Anderson et al., 2013.

Failed Forecast #3: A Thermal Hot Spot Should Exist in the Upper Troposphere in Tropical Regions

Observations from both weather balloon radiosondes and satellite MSU sensors show the opposite, with either flat or decreasing warming trends with increasing height in the troposphere (Douglass *et al.*, 2007; Singer, 2011; Singer, 2013a). In Figure 9, the image on the left is model simulations of temperature trends in the tropical mid-troposphere, as shown in figure 1.3F from a report by the U.S. Climate Change Science Program (Karl *et al.*, 2006). Figure on the right is figure 5.7E from the same source. These trends are based on the analysis of radiosonde data by the Hadley Centre and are in good agreement with the corresponding U.S. analyses. The observed data do not show the temperature rise in the tropical mid-troposphere forecast by the model.

Figure 9
Greenhouse-model-predicted Temperature Trends Versus Latitude and Altitude Versus Observed Temperature Trends

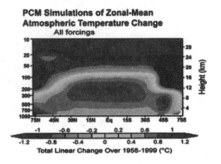
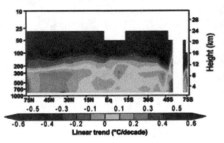

Source: Karl *et al.*, 2006, pp. 25, 116.

Failed Forecast #4: Both Polar Regions Should Have Warmed Faster than the Rest of Earth During the Late Twentieth Century

Late-twentieth century warming occurred in many Arctic locations and also over a limited area of the West Antarctic Peninsula, but the large polar East Antarctic Ice Sheet has been cooling since at least the 1950s (O'Donnell *et al.*, 2010). .

* * *

In general, GCMs perform poorly when their projections are assessed against empirical data. In their comprehensive report of an extensive test of contemporary climate models, Idso and Idso write, "we find (and document) a total of 2,418 failures of today's top-tier climate models to accurately hindcast a whole host of climatological phenomena. And with this extremely poor record of success, one must greatly wonder how it is that anyone would believe what the climate models of today project about earth's climate of tomorrow, i.e., a few decades to a century or more from now" (Idso and Idso, 2015).

References

Anderson, D., *et al.* 2013. Global warming in an independent record of the last 130 years. *Geophysical Research Letters* **40:** 189–193, doi:10.1029/2012GL054271.

Douglass, D.H., Christy, J.R., Pearson, B.D., and Singer, S.F. 2007. A comparison of tropical temperature trends with model predictions. *International Journal of Climatology* doi: 10.1002/joc.1651.

Dyson, F. 2004. A meeting with Enrico Fermi. *Nature* **427**: 297.

Gouretski, V.V., Kennedy, J. J. J., Boyer, T.P., and Köhl, A. 2012. Consistent near-surface ocean warming since 1900 in two largely independent observing networks, *Geophysical Research Letters*, doi:10.1029/2012GL052975.

Idso, S.B. and Idso, C.D. 2015. *Mathematical Models vs. Real-World Data: Which Best Predicts Earth's Climatic Future?* Center for the Study of Carbon

Dioxide and Global Change.

Karl, T.R., Hassol, S.J. , Miller, C.D., and Murray, W.L. 2006. (Eds.) *Temperature Trends in the Lower Atmosphere: Steps for Understanding and Reconciling Differences. A report by the Climate Change Science Program and Subcommittee on Global Change Research,* http://www.climatescience.gov/Library/sap/sap1-1/final report/default.htm.

Knight, J., Kennedy, J., Folland, C., Harris, G., Jones, G.S., Palmer, M., Parker, D., Scaife, A., and Stott, P. 2009. Do global temperature trends over the last decade falsify climate predictions? *Bulletin of the American Meteorological Society* **90:** S22–S23 (2009).

Lindzen, R.S. and Choi, Y.-S. 2011. On the observational determination of climate sensitivity and its implications. *Asia-Pacific Journal of Atmospheric Sciences* **47:** 377–390. doi:10.1007/s13143-011-0023-x.

Mears C.A. and Wentz, F.J. 2009. Construction of the RSS V3.2 lower tropospheric dataset from the MSU and AMSU microwave sounders. *Journal of Atmospheric and Oceanic Technology* **26:** 1493–1509.

Monckton, C., Soon, W.W.-H, Legates, D.R., and Briggs, W.M. 2015. Keeping it simple: the value of an irreducibly simple climate model. *Science Bulletin* **60:** 15, 1378–1390.

O'Donnell, R., Lewis, N., McIntyre, S., and Condon, J. 2010. Improved methods for PCA-based reconstructions: case study using the Steig *et al.* (2009) Antarctic temperature reconstruction. *Journal of Climate* **24:** 2099–2115.

Singer, S.F. 2011. Lack of consistency between modelled and observed temperature trends. *Energy & Environment* **22:** 375–406.

Singer, S.F. 2013a. Inconsistency of modelled and observed tropical temperature trends. *Energy & Environment* **24:** 405–413.

Singer, S.F. 2013b. Overcoming chaotic behavior of general circulation climate models (GCMs). *Energy & Environment* **24:** 397–403.

Spencer, R.W. and Braswell, W.D. 2008. Potential biases in feedback diagnosis from observations data: a simple model demonstration. *Journal of Climate* **21:** 5624–5628.

5

False Postulates

Key findings in this section include the following:

- Neither the rate nor the magnitude of the reported late twentieth century surface warming (1979–2000) lay outside normal natural variability.

- The late twentieth century warm peak was of no greater magnitude than previous peaks caused entirely by natural forcings and feedbacks.

- Historically, increases in atmospheric CO_2 followed increases in temperature, they did not precede them. Therefore, CO_2 levels could not have forced temperatures to rise.

- Solar forcings are not too small to explain twentieth century warming. In fact, their effect could be equal to or greater than the effect of CO_2 in the atmosphere.

- A warming of 2°C or more during the twenty-first century would probably not be harmful, on balance, because many areas of the world would benefit from or adjust to climate change.

Figure 2 in Chapter 3 identified five postulates at the base of IPCC's claim that global warming has resulted, or will result, from anthropogenic greenhouse gas emissions. All five are readily refuted by real-world observations.

Modern Warming Is Not Unnatural

IPCC's first false postulate is that the warming of the twentieth century cannot be explained by natural variability. But temperature records contain natural climate rhythms that are not well summarized or defined by fitting straight lines through arbitrary portions of a fundamentally rhythmic, non-stationary data plot. In particular, linear fitting fails to take account of meteorological-oceanographical-solar variations that are well established to occur at multidecadal and millennial time scales. Even assuming, wrongly, that global temperatures would have been unchanging in the absence of man-made greenhouse gas emissions, the correctness of IPCC's assertion depends upon the period of time considered (Davis and Bohling, 2001). For example, temperatures have been cooling since 8,000 and 2,000 years ago; warming since 20,000 years ago, and also since 1850 and since 1979; and static (no net warming or cooling) between 700 and 150 AD and since 1997 AD.

Global warming during the twentieth century occurred in two pulses, between 1910–1940 and 1975–2000, at gentle rates of a little over 1.5°C/century (British Meteorological Office, 2013). In contrast, natural warming at some individual meteorological stations during the 1920s proceeded at high rates of up to 4°C/decade or more (Chylek *et al.*, 2004). The first period (1910–1940) represents rates of global warming that are fully natural (having occurred prior to the major build-up of greenhouse gases in the atmosphere), whereas measurements made during the late twentieth century warming are likely exaggerated by inadequate correction for the urban heat island effect.

Comparison of modern and ancient rates of natural temperature change is difficult because of the lack of direct measurements available prior to 1850. However, high-quality proxy temperature records from the Greenland ice core for the past 10,000 years demonstrate a natural range of warming and cooling rates between +2.5 and -2.5 °C/century (Alley, 2000; Carter, 2010, p. 46, figure 7), significantly greater than rates measured for Greenland or the globe during the twentieth century.

Modern Warming Is Not Unprecedented

IPCC's second false postulate is that the late twentieth century warm peak was of greater magnitude than previous natural peaks. But the glaciological and recent geological records contain numerous examples of ancient temperatures up to 3°C or more warmer than the peak reported at the end of the twentieth century. During the Holocene, such warmer peaks included the Egyptian, Minoan, Roman, and Medieval warm periods (Alley, 2000). During the Pleistocene, warmer peaks were associated with interglacial oxygen isotope stages 5, 9, 11, and 31 (Lisiecki and Raymo, 2005). During the Late Miocene and Early Pliocene (6–3 million years ago) temperature consistently attained values 2–3°C above twentieth century values (Zachos *et al.*, 2001).

Figure 10 summarizes these and other findings about surface temperatures that appear in Chapter 4 of *Climate Change Reconsidered-II: Physical Science.*

Figure 10
Key Facts about Surface Temperature

- Whether today's global surface temperature is seen to be part of a warming trend depends upon the time period considered.

- Over (climatic) time scales of many thousand years, temperature is cooling; over the historical (meteorological) time scale of the past century temperature has warmed. Over the past 18 years, there has been no net warming despite an increase in atmospheric CO_2 of 8 percent – which represents 34 percent of all human-related CO_2 emissions released to the atmosphere since the industrial revolution.

- Given an atmospheric mixing time of ~1 year, the facts just related represent a test of the dangerous warming hypothesis, which test it fails.

- Based upon the HadCRUT dataset favored by IPCC, two phases of warming occurred during the twentieth century, between 1910–1940 and 1979–2000, at similar rates of a little over 1.5°C/century. The early twentieth century warming preceded major industrial carbon dioxide emissions and must be natural; warming during the second (*prima facie,* similar) period might incorporate a small human-related carbon dioxide effect, but warming might also be inflated by urban heat island effects.

- Other temperature datasets fail to record the late twentieth century warming seen in the HadCRUT dataset.

- There was nothing unusual about either the magnitude or rate of the late twentieth century warming pulses represented on the HadCRUT record, both falling well within the envelope of known, previous natural variations.

- No empirical evidence exists to support the assertion that a planetary warming of 2°C would be net ecologically or economically damaging.

Source: "Chapter 4. Observations: Temperatures," *Climate Change Reconsidered II: Physical Science* (Chicago, IL: The Heartland Institute, 2013).

CO$_2$ Does Not Lead Temperature

IPCC's third false postulate is that increases in atmospheric CO$_2$ precede, and then force, parallel increases in temperature. The remarkable (and at first blush, synchronous) parallelism that exists between rhythmic fluctuations in ancient atmospheric temperature and atmospheric CO$_2$ levels was first detected in polar ice core samples analyzed during the 1970s. From the early 1990s onward, however, higher-resolution sampling has repeatedly shown these historic temperature changes precede the parallel changes in CO$_2$ by several hundred years or more (Mudelsee, 2001). A similar relationship of temperature change leading CO$_2$ change (in this case by several months) also characterizes the much shorter seasonal cyclicity manifest in Hawaiian and other meteorological measurements (Kuo *et al.*, 1990).

In such circumstances, changing levels of CO$_2$ cannot be driving changes in temperature, but must either be themselves stimulated by temperature change, or be co-varying with temperature in response to changes in another (at this stage unknown) variable.

Solar Influence Is Not Minimal

IPCC's fourth false postulate is that solar forcings are too small to explain twentieth century warming. Having concluded solar forcing alone is inadequate to account for twentieth century warming, IPCC authors infer CO_2 must be responsible for the remainder. Nonetheless, observations indicate variations occur in total ocean–atmospheric meridional heat transport and that these variations are driven by changes in solar radiation rooted in the intrinsic variability of the Sun's magnetic activity (Soon and Legates, 2013).

Incoming solar radiation is most often expressed as Total Solar Insolation (TSI), a measure derived from multi-proxy measures of solar activity (Hoyt and Schatten, 1993; extended and re-scaled by Willson, 2011; Scafetta and Willson, 2013). The newest estimates, from satellite-borne ACRIM-3 measurements, indicate TSI ranged between 1360 and 1363 Wm-2 between 1979 and 2011, the variability of ~3 Wm-2 occurring in parallel with the 11-year sunspot cycle. Larger changes in TSI are also known to occur in parallel with climatic change over longer time scales. For instance, Shapiro *et al.* (2011) estimated the TSI change between the Maunder Minimum and current conditions may have been as large as 6 Wm-2.

Temperature records from circum-Arctic regions of the Northern Hemisphere show a close correlation with TSI over the past 150 years, with both measures conforming to the ~60–70 year multidecadal cycle. In contrast, the measured steady rise of CO_2 emissions over the same period shows little correlation with the strong multidecadal (and shorter) ups and downs of surface temperature around the world.

Finally, IPCC ignores x-ray, ultraviolet, and magnetic flux variation, the latter having particularly important implications for the modulation of galactic cosmic ray influx and low cloud formation (Svensmark, 1998; Kirkby, *et al.*, 2011). Figure 11 summarizes these and other findings about solar forcings from Chapter 3 of *Climate Change Reconsidered II: Physical Science.*

Figure 11
Key Facts about Solar Forcing

■ Evidence is accruing that changes in Earth's surface temperature are largely driven by variations in solar activity. Examples of solar-controlled climate change epochs include the Medieval Warm Period, Little Ice Age, and Early Twentieth Century (1910–1940) Warm Period.

■ The Sun may have contributed as much as 66 percent of the observed twentieth century warming, and perhaps more.

■ Strong empirical correlations have been reported from around the world between solar variability and climate indices including temperature, precipitation, droughts, floods, streamflow, and monsoons.

■ IPCC models do not incorporate important solar factors such as fluctuations in magnetic intensity and overestimate the role of human-related CO_2 forcing.

■ IPCC fails to consider the importance of the demonstrated empirical relationship between solar activity, the ingress of galactic cosmic rays, and the formation of low clouds.

■ The respective importance of the Sun and CO_2 in forcing Earth's climate remains unresolved; current climate models fail to account for a plethora of known Sun-climate connections.

■ The recently quiet Sun and extrapolation of solar cycle patterns into the future suggest a planetary cooling may occur over the next few decades.

Source: "Chapter 3. Solar Forcing of Climate," *Climate Change Reconsidered II: Physical Science* (Chicago, IL: The Heartland Institute, 2013).

Warming Would Not Be Harmful

IPCC's fifth false postulate is that warming of 2°C above today's temperature would be harmful. The suggestion that 2°C of warming would be harmful was coined at a conference organized by the British Meteorological Office in 2005 (DEFRA, 2005). The particular value of 2°C is entirely arbitrary and was proposed by the World Wildlife Fund, an environmental advocacy group, as a political expediency rather than as an informed scientific opinion. The target was set in response to concern that politicians would not initiate policy actions to reduce CO_2 emissions unless they were given a specific (and low) quantitative temperature target to aim for.

Multiple lines of evidence suggest a 2°C rise in temperature would not be harmful to the biosphere. The period termed the Holocene Climatic Optimum (c. 8,000 ybp) was 2–3°C warmer than today (Alley, 2000), and the planet attained similar temperatures for several million years during the Miocene and Pliocene (Zachos *et al.*, 2001). Biodiversity is encouraged by warmer rather than colder temperatures (Idso and Idso, 2009), and higher temperatures and elevated CO_2 greatly stimulate the growth of most plants (Idso and Idso, 2011).

Despite its widespread adoption by environmental NGOs, lobbyists, and governments, no empirical evidence exists to substantiate the claim that 2°C of warming presents a threat to planetary ecologies or human well-being. Nor can any convincing case be made that a warming will be more economically costly than an equivalent cooling (either of which could occur unheralded for entirely natural reasons), since any planetary change of 2°C magnitude in temperature would result in complex local and regional changes, some being of economic or environmental benefit and others being harmful.

We conclude neither the rate nor the magnitude of the reported late twentieth century surface warming (1979–2000) lay outside normal natural variability, nor was it in any way unusual compared to earlier episodes in Earth's climatic history. Furthermore, solar forcings of temperature change are likely more important than is currently recognized, and evidence is lacking that a 2°C increase in temperature (of whatever cause) would be globally harmful.

References

Alley, R.B. 2000. The Younger Dryas cold interval as viewed from central Greenland. *Quaternary Science Reviews* **19**: 213–226.

British Meteorological Office. 2013. Met Office Hadley Centre observations datasets. CRUTEM4 Data. http://www.metoffice.gov.uk/hadobs/ crutem4/data/download.html.

Carter, R.M. 2010. *Climate: The Counter Consensus.* London, UK: Stacey International.

Chylek, P., Figure, J.E., and Lesins, G. 2004. Global warming and the Greenland ice sheet. *Climatic Change* **63**: 201–221.

Davis, J.C. and Bohling, G.C. 2001. The search for pattern in ice-core temperature curves. American Association of Petroleum Geologists, *Studies in Geology* **47**: 213–229.

DEFRA 2005. Symposium on avoiding dangerous climate change. Exeter, Feb. 1–3, http://www.stabilisation2005.com/.

Hoyt, D.V. and Schatten, K.H. 1993. A discussion of plausible solar irradiance variations, 1700–1992. *Journal of Geophysical Research* **98**: 18895–18906. http://dx.doi.org/10.1029/93JA01944.

Idso, C.D. and Idso, S.B. 2009. *CO₂, Global Warming and Species Extinctions: Prospects for the Future.* Pueblo West, CO: Vales Lake Publishing.

Idso, C.D. and Idso, S.B. 2011. *The Many Benefits of Atmospheric CO₂ Enrichment.* Pueblo West, CO: Vales Lake Publishing.

Kirkby, J. *et al.* 2011. Role of sulphuric acid, ammonia and galactic cosmic rays in atmospheric aerosol nucleation. *Nature* **476**: 429–433.

Kuo, C., Lindberg, C., and Thomson, D.J. 1990. Coherence established between atmospheric carbon dioxide and global temperature. *Nature* **343**: 709–713.

Lisiecki, L.E. and Raymo, M.E. 2005. A Pliocene-Pleistocene stack of 57 globally distributed benthic d¹⁸O records. *Paleoceanography* **20**: PA1003. doi:10.1029/2004PA001071.

Mudelsee, M. 2001. The phase relations among atmospheric CO_2 content, temperature and global ice volume over the past 420 ka. *Quaternary Science Reviews* **20**: 583–589.

Scafetta, N. and Willson, R.C. 2013. Empirical evidences for a planetary modulation of total solar irradiance and the TSI signature of the 1.09-year Earth-Jupiter conjunction cycle. *Astrophysics and Space Sciences*, doi:10.1007/s10509-013-1558-3.

Shapiro, A.I., Schmutz, W., Rozanov, E., Schoell, M., Haberreiter, M., Shapiro, A.V., and Nyeki, S. 2011. A new approach to the long-term reconstruction of the solar irradiance leads to a large historical solar forcing. *Astronomy and Astrophysics* **529**: A67.

Soon, W. and Legates, D.R. 2013. Solar irradiance modulation of equator-to-pole (Arctic) temperature gradients: Empirical evidence for climate variation on multi-decadal timescales. *Journal of Atmospheric and Solar-Terrestrial Physics* **93**: 45–56.

Svensmark, H. 1998. Influence of cosmic rays on Earth's climate. *Physical Review Letters* **22**: 5027–5030.

Willson, R.C. 2011. Revision of ACRIMSAT/ACRIM3 TSI results based on LASP/TRF diagnostic test results for the effects of scattering, diffraction and basic SI scale traceability. Abstract for 2011 Fall AGU Meeting (Session GC21).

Zachos, J., Pagani, M., Sloan, L., Thomas, E., and Billups, K. 2001. Trends, rhythms, and aberrations in global climate 65 Ma to present. *Science* **292**: 686–693.

6

Unreliable Circumstantial Evidence

Key points in this chapter include the following:

■ Melting of Arctic sea ice and polar icecaps is not occurring at "unnatural" rates and does not constitute evidence of a human impact on climate.

■ Best available data show sea-level rise is not accelerating. Local and regional sea levels continue to exhibit typical natural variability – in some places rising and in others falling.

■ The link between warming and drought is weak, and by some measures drought has decreased over the twentieth century. Changes in the hydrosphere of this type are regionally highly variable and show a closer correlation with multidecadal climate rhythmicity than they do with global temperature.

■ No convincing relationship has been established between warming over the past 100 years and increases in extreme weather events. Meteorological science suggests just the opposite: A warmer world will see more mild weather patterns.

■ No evidence exists that current changes in Arctic permafrost are other than natural or are likely to cause a climate catastrophe by releasing methane into the atmosphere.

Introduction

IPCC's third line of reasoning, summarized in Figure 2 in Chapter 3, consists of circumstantial evidence regarding natural phenomena known to vary with temperature. The examples IPCC chooses to report invariably point to a negative impact on plant and animal life and human well-being. When claims are made that such phenomena are the result of anthropogenic global warming, almost invariably at least one of the following three requirements of scientific confidence is lacking:

(1) *Correlation does not establish causation.* Correlation of, say, a declining number of polar bears and a rising temperature does not establish causation between one and the other, for it is not at all unusual for two things to co-vary in parallel with other forcing factors.

(2) *Control for natural variability.* We live on a dynamic planet in which all aspects of the physical and biological environment are in a constant state of flux for reasons that are entirely natural (including, of course, temperature change). It is wrong to assume no changes would occur in the absence of the human presence. Climate, for example, will be different in 100 years regardless of what humans do or don't do.

(3) *Local temperature records that confirm warming.* Many studies of the impact of climate change on wildlife simply assume temperatures have risen, extreme weather events are more frequent, etc., without establishing that the relevant local temperature records conform to the postulated simple long-term warming trend.

All five of IPCC's claims relying on circumstantial evidence listed in Figure 2 in Chapter 3 are refutable.

Melting Ice

IPCC claims unusual melting is occurring in mountain glaciers, Arctic sea ice, and polar icecaps. But what melting is occurring in mountain glaciers, Arctic sea ice, and polar icecaps is not occurring at "unnatural" rates and does not constitute evidence of a human impact on the climate. Both the Greenland (Johannessen *et al.*, 2005; Zwally *et al.*, 2005) and Antarctic (Zwally and Giovinetto, 2011) icecaps are close to balance. The global area of sea ice today is similar to that first measured by satellite observation in 1979 (Humlum, 2013) and significantly exceeds the ice cover present in former, warmer times.

Valley glaciers wax and wane on multidecadal, centennial, and millennial time-scales, and no evidence exists that their present, varied behavior falls outside long-term norms or is related to human-related CO_2 emissions (Easterbrook, 2011). Figure 12 summarizes the findings of Chapter 5 of *Climate Change Reconsidered-II: Physical Science* regarding glaciers, sea ice, and polar icecaps.

Figure 12
Key Facts about the Cryosphere

- Satellite and airborne geophysical datasets used to quantify the global ice budget are short and the methods involved in their infancy, but results to date suggest both the Greenland and Antarctic Ice Caps are close to balance.

- Deep ice cores from Antarctica and Greenland show climate change occurs as both major glacial-interglacial cycles and as shorter decadal and centennial events with high rates of warming and cooling, including abrupt temperature steps.

- Observed changes in temperature, snowfall, ice flow speed, glacial extent, and iceberg calving in both Greenland and Antarctica appear to lie within the limits of natural climate variation.

- Global sea-ice cover remains similar in area to that at the start of satellite observations in 1979, with ice shrinkage in the Arctic Ocean since then being offset by growth around Antarctica.

- During the past 25,000 years (late Pleistocene and Holocene) glaciers around the world have fluctuated broadly in concert with changing climate, at times shrinking to positions and volumes smaller than today.

- This fact notwithstanding, mountain glaciers around the world show a wide variety of responses to local climate variation and do not respond to global temperature change in a simple, uniform way.

- Tropical mountain glaciers in both South America and Africa have retreated in the past 100 years because of reduced precipitation and increased solar radiation; some glaciers elsewhere also have retreated since the end of the Little Ice Age.

- The data on global glacial history and ice mass balance do not support the claims made by IPCC that CO_2 emissions are causing most glaciers today to retreat and melt.

Source: "Chapter 5. Observations: The Cryosphere," *Climate Change Reconsidered II: Physical Science* (Chicago, IL: The Heartland Institute, 2013).

Sea-Level Rise

IPCC claims global sea level is rising at an enhanced rate and swamping tropical coral atolls. But the best available data show sea-level rise is not accelerating (Houston and Dean, 2011). The global average sea level continues to increase at its long-term rate of 1–2 mm/year globally (Wöppelmann *et al.*, 2009). Local and regional sea levels continue to exhibit typical natural variability – in some places rising and in others falling. Unusual sea-level rise is therefore not drowning Pacific coral islands, nor are the islands being abandoned by "climate refugees."

The best available data show dynamic variations in Pacific sea level vary in accord with El Niño-La Niña cycles, superimposed on a natural long-term eustatic rise (Australian Bureau of Meteorology, 2011). Island coastal flooding results not from sea-level rise, but from spring tides or storm surges in combination with development pressures such as borrow pit digging or groundwater withdrawal. Persons emigrating from the islands are doing so for social and economic reasons rather than in response to environmental threat.

Another claim concerning the effect of climate change on oceans is that increases in freshwater runoff into the oceans will disrupt the global thermohaline circulation system. But the range of natural fluctuation in the global ocean circulation system has yet to be fully delineated (Srokosz *et al.*, 2012). Research to date shows no evidence for changes that lie outside previous natural variability, nor for any malign influence from increases in human-related CO_2 emissions. See Figure 13 for more findings about climate change and oceans from Chapter 6 of *Climate Change Reconsidered II: Physical Science.*

Figure 13
Key Facts about Oceans

- Knowledge of local sea-level change is vital for coastal management; such change occurs at widely variable rates around the world, typically between about +5 and -5 mm/year.

- Global (eustatic) sea level, knowledge of which has only limited use for coastal management, rose at an average rate of between 1 and 2 mm/year over the past century.

- Satellite altimeter studies of sea-level change indicate rates of global rise since 1993 of more than 3 mm/year, but complexities of processing and the infancy of the method preclude viewing this result as secure.

- Rates of global sea-level change vary in decadal and multidecadal ways and show neither recent acceleration nor any simple relationship with increasing CO_2 emissions.

- Pacific coral atolls are not being drowned by extra sea-level rise; rather, atoll shorelines are affected by direct weather and infrequent high tide events, ENSO sea-level variations, and impacts of increasing human populations.

- Extra sea-level rise due to heat expansion (thermosteric rise) is also unlikely given that the Argo buoy network shows no significant ocean warming over the past nine years (Knox and Douglass, 2010).

- Though the range of natural variation has yet to be fully described, evidence is lacking for any recent changes in global ocean circulation that lie outside natural variation or were forced by human CO_2 emissions.

Source: "Chapter 6. Observations: The Hydrosphere," *Climate Change Reconsidered II: Physical Science* (Chicago, IL: The Heartland Institute, 2013).

Droughts, Floods, and Monsoons

IPCC claims droughts, floods, and monsoon variability and intensity are increasing. But the link between warming and drought is weak, and pan evaporation (a measurement that responds to the effects of several climate elements) decreased over the twentieth century (Roderick *et al.*, 2009). Huntington (2008) concluded on a globally averaged basis precipitation over land increased by about 2 percent over the period 1900–1998. However, changes in the hydrosphere of this type are regionally highly variable and show a closer correlation with multidecadal climate rhythmicity than they do with global temperature (Zanchettin *et al.*, 2008).

Monsoon intensity correlates with variations in solar activity rather than increases in atmospheric CO_2, and both the South American and Asian monsoons became more active during the cold Little Ice Age and less active during the Medieval Warm Period (Vuille *et al.*, 2012), suggesting there would be less volatility if the world becomes warmer. See Figure 14 for more facts about monsoons, droughts, and floods presented in Chapter 6 of *Climate Change Reconsidered II: Physical Science.*

Figure 14
Key Facts about Monsoons, Droughts, and Floods

■ Little evidence exists for an overall increase in global precipitation during the twentieth century independent of natural multidecadal climate rhythmicity.

■ Monsoon precipitation did not become more variable or intense during late twentieth century warming; instead, precipitation responded mostly to variations in solar activity.

■ South American and Asian monsoons were more active during the cold Little Ice Age and less active during the Medieval Warm Period. Neither global nor local changes in streamflow have been linked to CO_2 emissions.

■ The relationship between drought and global warming is weak, since severe droughts occurred during both the Medieval Warm Period and the Little Ice Age.

Source: "Chapter 6. Observations: The Hydrosphere," *Climate Change Reconsidered II: Physical Science* (Chicago, IL: The Heartland Institute, 2013).

Extreme Weather

IPCC does not object when persons, such as former U.S. Vice President Al Gore, cite its reports in support of claims that global warming is leading to more, or more intense, wildfires, rainfall, storms, hurricanes, and other extreme weather events. IPCC's latest Summary for Policymakers is filled with vivid warnings of this kind, even though in 2012 an IPCC report acknowledged that a relationship between global warming and wildfires, rainfall, storms, hurricanes, and other extreme weather events has not been demonstrated (IPCC, 2012).

In no case has a convincing relationship been established between warming over the past 100 years and increases in any of these extreme weather events (Pielke, Jr., 2014). Instead, the number and intensity of extreme events vary, and they wax and wane from one place to another and often in parallel with natural decadal or multidecadal climate oscillations. Basic meteorological science suggests a warmer world would experience fewer storms and weather extremes, as indeed has been the case in recent years. Figure 15 summarizes key facts on this subject presented in Chapter 7 of *Climate Change Reconsidered-II: Physical Science.*

Figure 15
Key Facts about Extreme Weather Events

- Air temperature variability decreases as mean air temperature rises, on all time scales.

- Therefore the claim that global warming will lead to more extremes of climate and weather, including of temperature itself, seems theoretically unsound; the claim is also unsupported by empirical evidence.

- Although specific regions have experienced significant changes in the intensity or number of extreme events over the twentieth century, for the globe as a whole no relationship exists between such events and global warming over the past 100 years.

- Observations from across the planet demonstrate that droughts have not become more extreme or erratic in response to global warming. In most cases, the worst droughts in recorded meteorological history were much milder than droughts that occurred periodically during much colder times.

- There is little to no evidence that precipitation will become more variable and intense in a warming world; indeed some observations show just the opposite.

- There has been no significant increase in either the frequency or intensity of stormy weather in the modern era.

- Despite the supposedly "unprecedented" warming of the twentieth century, there has been no increase in the intensity or frequency of tropical cyclones globally or in any of the specific ocean basins.

- The commonly held perception that twentieth century warming was accompanied by an increase in extreme weather events is a misconception fostered by excessive media attention and has no basis in facts (Khandekar, 2013).

Source: "Chapter 7. Observations: Extreme Weather," *Climate Change Reconsidered II: Physical Science* (Chicago, IL: The Heartland Institute, 2013).

Thawing Permafrost

IPCC claims unusual thawing of Boreal permafrost or sub-seabed gas hydrates is causing warming due to methane release. It is true that over historic time, methane concentration has increased from about 700 ppb in the eighteenth century to the current level of near 1,800 ppb. However, the increase in methane concentration levelled off between 1998 and 2006 at around 1,750 ppb, which may reflect measures taken at that time to stem leakage from wells, pipelines, and distribution facilities (Quirk, 2010). More recently, since about 2007, methane concentrations have started to increase again, possibly due to a combination of leaks from new shale gas drilling and Arctic permafrost decline.

The contribution of increased methane to radiation forcing since the eighteenth century is estimated to be only 0.7 Wm^{-2}, which is small. And in any case, no evidence exists that current changes in Arctic permafrost are other than natural. Most of Earth's gas hydrates occur at low saturations and in sediments at such great depths below the seafloor or onshore permafrost that they will barely be affected by warming over even one thousand years.

* * *

We conclude no unambiguous evidence exists for adverse changes to the global environment caused by human-related CO_2 emissions. In particular, the cryosphere is not melting at an enhanced rate; sea-level rise is not accelerating; no systematic changes have been documented in evaporation or rainfall or in the magnitude or intensity of extreme meteorological events; and an increased release of methane into the atmosphere from permafrost or sub-seabed gas hydrates is unlikely.

References

Australian Bureau of Meteorology. 2011. The South Pacific sea-level and climate monitoring program. Sea-level summary data report, July 2010–June 2010. http://www.bom.gov.au/ntc/IDO60102/IDO60102.2011_1.pdf.

Easterbrook, D.J. (Ed.) 2011. *Evidence-based Climate Science*. Amsterdam: Elsevier Inc.

Houston, J.R. and Dean, R.G. 2011. Sea-level acceleration based on U.S. tide gauges and extensions of previous global-gauge analyses. *Journal of Coastal Research* **27:** 409–417.

Humlum, O. 2013. Monthly Antarctic, Arctic and global sea ice extent since November 1978, after National Snow and Ice Data Center, USA. http://www.climate4you.com/.

Huntington, T.G. 2008. Can we dismiss the effect of changes in land-based water storage on sea-level rise? *Hydrological Processes* **22:** 717–723.

Intergovernmental Panel on Climate Change. 2012. Special Report on Managing the Risks of Extreme Events and Disasters to Advance Climate Change Adaptation (SREX). http://ipcc wg2.gov/SREX/report/.

Johannessen, O.M., Khvorostovsky, K., Miles, M.W., and Bobylev, L.P. 2005. Recent ice-sheet growth in the interior of Greenland. *Science* **310:** 1013–1016.

Khandekar, M.L. 2013. Are extreme weather events on the rise? *Energy & Environment* **24:** 537–549.

Knox, R.S. and Douglass, D.H. 2010. Recent energy balance of Earth. *International Journal of Geosciences* 1. doi:10.4236/ijg2010.00000.

Pielke Jr., R.A. 2014. *The Rightful Place of Science: Disasters and Climate Change.* Tempe, AZ: Arizona State University Consortium for Science, Policy & Outcomes.

Quirk, T. 2010. Twentieth century sources of methane in the atmosphere. *Energy & Environment* 21: 251–266.

Roderick, M.L., Hobbins, M.T., and Farquhar, G.D. 2009. Pan evaporation trends and the terrestrial water balance. II. Energy balance and interpretation. *Geography Compass* 3: 761–780, doi: 10.1111/j.1749-8198.2008.0021.

Srokosz, M., Baringer, M., Bryden, H., Cunningham, S., Delworth, T., Lozier, S., Marotzke, J., and Sutton, R. 2012. Past, present, and future changes in the Atlantic Meridional Overturning Circulation. *Bulletin of the American Meteorological Society* 93: 1663–1676. doi:10.1175/BAMS-D-11-00151.1.

Vuille, M., Burns, S.J., Taylor, B.L., Cruz, F.W., Bird, B.W., Abbott, M.B., Kanner, L.C., Cheng, H., and Novello, V.F. 2012. A review of the South American monsoon history as recorded in stable isotopic proxies over the past two millennia. *Climate of the Past* 8: 1309–1321.

Wöppelmann, G., Letetrel, C., Santamaria, A., Bouin, M.-N., Collilieux, X., Altamimi, Z., Williams, S.D.P., and Miguez, B.M. 2009. Rates of sea-level change over the past century in a geocentric reference frame. *Geophysical Research Letters* 36: 10.1029/2009GL038720.

Zanchettin, D., Franks, S.W., Traverso, P., and Tomasino, M. 2008. On ENSO impacts on European wintertime rainfalls and their modulation by the NAO and the Pacific multi-decadal variability. *International Journal of Climatology* 28: 1995–1006. http://dx.doi.org/10.1002/joc.1601.

Zwally, H.J. and Giovinetto, M.B. 2011. Overview and assessment of Antarctic Ice-Sheet mass balance estimates: 1992–2009. *Surveys in Geophysics* 32: 351–376.

Zwally, H.J., Giovinetto, M.B., Li, J., Cornejo, H.G., Beckley, M.A., Brenner, A.C., Saba, J.L., and Yi, D. 2005. Mass changes of the Greenland and Antarctic ice sheets and shelves and contributions to sea-level rise: 1992–2002. *Journal of Glaciology* 51: 509–527.

7

Policy Implications

Key findings in this section include the following:

- Rather than rely exclusively on IPCC for scientific advice, policymakers should seek out advice from independent, nongovernment organizations and scientists who are free of financial and political conflicts of interest.

- Individual nations should take charge of setting their own climate policies based upon the hazards that apply to their particular geography, geology, weather, and culture.

- Rather than invest scarce world resources in a quixotic campaign based on politicized and unreliable science, world leaders would do well to turn their attention to the real problems their people and their planet face.

To date, most government signatories to the UN's Framework Convention on Climate Change have deferred to the monopoly advice of IPCC in setting their national climate change policies. More than 20 years since IPCC began its work in 1988, it is now evident this approach has been mistaken. One result has been the expenditure of hundreds of billions of dollars implementing energy policies that now appear to have been unnecessary, or at least ill-timed and ineffective.

Rather than rely exclusively on IPCC for scientific advice,

policymakers should seek out advice from independent, nongovernment organizations and scientists who are free of financial and political conflicts of interest. The Chinese Academy of Sciences took an important step in this direction by translating and publishing an abridged edition of the first two volumes in NIPCC's Climate Change Reconsidered series (CAS, 2013).

Climate change, whether man-made or not, is a global phenomenon with very different effects on different parts of the world (Tol, 2011). Individual nations should take charge of setting their own climate policies based upon the hazards that apply to their particular geography, geology, weather, and culture – as India has started to do by setting up an advisory Indian Network on Comprehensive Climate Change Assessment (INCCCA) (Nelson, 2010).

The theoretical hazard of dangerous human-caused global warming is but one small part of a much wider climate hazard – extreme natural weather and climatic events that Nature intermittently presents us with, and always will (Carter, 2010). The 2005 Hurricane Katrina disaster in the United States, the 2007 floods in the United Kingdom, and the tragic bushfires in Australia in 2009 demonstrate the governments of even advanced, wealthy countries are often inadequately prepared for climate-related disasters of natural origin.

Climate change as a natural hazard is as much a geological as a meteorological issue. Geological hazards are mostly dealt with by providing civil defense authorities and the public with accurate, evidence-based information regarding events such as earthquakes, volcanic eruptions, tsunamis, storms, and floods (which represent climatic as well as weather events), and then planning to mitigate and adapt to the effects when such events occur.

The idea that there can be a one-size-fits-all global solution to address future climate change, such as recommended by the United Nations in the past, fails to deal with real climate and climate-related hazards. It also turned climate change into a political issue long before the science was sufficiently advanced to inform policymakers. A better path forward was suggested by Ronald Brunner and Amanda Lynch: "We need to use adaptive governance to produce response programs that cope with hazardous climate events as they happen, and that encourage diversity and innovation in the search for solutions. In such a fashion, the highly contentious "global warming" problem can be recast into an issue in which every culture and community around the world has an inherent interest"

(Brunner and Lynch, 2010).

There is some evidence world leaders are reconsidering past decisions. India, China, Russia, and other countries are making it clear they will not blindly follow the path of reducing the use of fossil fuels in the vain hope of having an almost indiscernible effect on climate some time in the twenty-second or twenty-third centuries. A writer for *Nature*, commenting on the upcoming Conference of the Parties (COP-21) of the UN Framework Convention on Climate Change, reported in May 2015, "The negotiations' goal has become what is politically possible, not what is environmentally desirable. Gone is a focus on establishing a global, 'top down' target for stabilizing emissions of a carbon budget that is legally binding. The Paris meeting will focus on voluntary 'bottom up' commitments by individual states to reduce emissions. The global climate target is being watered down in the hope of getting any agreement in Paris. The 2°C warming limit need only be kept 'within reach.' The possibility of using 'ratcheting mechanisms' keeps hopes alive of more ambitious policies, but such systems are unlikely to achieve the desired outcomes. Strict measuring, reporting and verification mechanisms are yet to be agreed" (Geden, 2015, p. 27).

Michael Levi, a senior fellow for the Council on Foreign Relations, wrote in June 2015 about the changing expectations of world leaders. His points in brief: (1) Developed countries are no longer pushing for binding emissions reduction commitments, whether for themselves or developing countries; (2) the emphasis has shifted from reducing emissions in order to mitigate future climate change to helping nations adapt to whatever the future climate might look like; (3) the goals declared at the UN's next meeting (in Paris in December 2015) will be too far in the future to matter to anyone; and (4) the widely discussed pledge of giving developing countries $100 billion a year is going to consist largely of relabeling foreign aid and private funding already going to those countries (Levi, 2015).

If Geden's and Levi's observations are true, this is all very good news indeed. The world appears to be backing away from a disaster of its own making, caused by lobbyists and campaigners and interest groups steering public policy in the wrong direction.

Policymakers should recognize that the human impact on the global climate remains a scientific puzzle, perhaps the most difficult one science has ever faced. The scientific debate is far from over. Despite appeals to a "scientific consensus" and claims from even the president of the United

States that "climate change is real, man-made, and dangerous," the truth is we simply don't know if climate change is a problem that needs to be addressed. The best available evidence points in a different direction: The human impact on climate is small relative to natural variability, perhaps too small to be measured. Rather than invest scarce world resources in a quixotic campaign based on politicized and unreliable science, world leaders would do well to turn their attention to the real problems their people and their planet face.

References

Brunner, R.D. and Lynch, A.H. 2010. *Adaptive Governance and Climate Change*. Boston, MA: Meteorological Society of America. ISBN 9781878220974.

Carter, R.M. 2010. *Climate: The Counter Consensus*. London, UK: Stacey International.

CAS. 2013. Chinese Academy of Science. *Climate Change Reconsidered*. Chinese language edition translated by China Information Center for Global Change Studies, www.globalchange.ac.cn, and published by Science Press. ISBN 978-7-03-037484-4.

Geden, O. 2015. Climate advisers must maintain integrity. *Nature* **521**: (May 7) 27–28.

Levi, M. 2015. What matters (and what doesn't) in the G7 climate declaration. Website, Council on Foreign Relations (June 10), http://blogs.cfr.org/levi/2015/06/10/what-matters-and-what-doesnt-in-the-g7-climate-declaration/. Last viewed on October 30, 2015.

Nelson, D. 2010. India forms new climate change body. Feb. 4, 2010. The Telegraph (UK). http://www.telegraph.co.uk/earth/environment/climatechange/7157590/India-forms-new-climate-change-body.html.

Conclusion

The most important fact about climate science, often overlooked, is that scientists disagree about the environmental impacts of the combustion of fossil fuels on the global climate. There is no survey or study showing "consensus" on the most important scientific issues, despite frequent claims by advocates to the contrary.

Scientists disagree about the causes and consequences of climate for several reasons. Climate is an interdisciplinary subject requiring insights from many fields. Very few scholars have mastery of more than one or two of these disciplines. Fundamental uncertainties arise from insufficient observational evidence, disagreements over how to interpret data, and how to set the parameters of models. The Intergovernmental Panel on Climate Change (IPCC), created to find and disseminate research finding a human impact on global climate, is not a credible source. It is agenda-driven, a political rather than scientific body, and some allege it is corrupt. Finally, climate scientists, like all humans, can be biased. Origins of bias include careerism, grant-seeking, political views, and confirmation bias.

Probably the only "consensus" among climate scientists is that human activities can have an effect on local climate and that the sum of such local effects could hypothetically rise to the level of an observable global signal. The key questions to be answered, however, are whether the human global signal is large enough to be measured and if it is, does it represent, or is it likely to become, a dangerous change outside the range of natural variability? On these questions, an energetic scientific debate is taking place on the pages of peer-reviewed science journals.

In contradiction of the scientific method, IPCC assumes its implicit hypothesis – that dangerous global warming is resulting, or will result, from human-related greenhouse gas emissions – is correct and that its only duty is to collect evidence and make plausible arguments in the hypothesis's favor. It simply ignores the alternative and null hypothesis, amply supported by empirical research, that currently observed changes in global climate indices and the physical environment are the result of natural variability.

The results of the global climate models (GCMs) relied on by IPCC are only as reliable as the data and theories "fed" into them. Most climate scientists agree those data are seriously deficient and IPCC's estimate for climate sensitivity to CO_2 is too high. We estimate a doubling of CO_2 from pre-industrial levels (from 280 to 560 ppm) would likely produce a temperature forcing of 3.7 Wm^{-2} in the lower atmosphere, for about ~1°C of *prima facie* warming. The recently quiet Sun and extrapolation of solar cycle patterns into the future suggest a planetary cooling may occur over the next few decades.

In a similar fashion, all five of IPCC's postulates, or assumptions, are readily refuted by real-world observations, and all five of IPCC's claims relying on circumstantial evidence are refutable. For example, in contrast to IPCC's alarmism, we find neither the rate nor the magnitude of the reported late twentieth century surface warming (1979–2000) lay outside normal natural variability, nor was it in any way unusual compared to earlier episodes in Earth's climatic history. In any case, such evidence cannot be invoked to "prove" a hypothesis, but only to disprove one. IPCC has failed to refute the null hypothesis that currently observed changes in global climate indices and the physical environment are the result of natural variability.

Rather than rely exclusively on IPCC for scientific advice, policymakers should seek out advice from independent, nongovernment organizations and scientists who are free of financial and political conflicts of interest. NIPCC's conclusion, drawn from its extensive review of the scientific evidence, is that any human global climate impact is within the background variability of the natural climate system and is not dangerous.

In the face of such facts, the most prudent climate policy is to prepare for and adapt to extreme climate events and changes regardless of their origin. Adaptive planning for future hazardous climate events and change should be tailored to provide responses to the known rates, magnitudes, and risks of natural change. Once in place, these same plans will provide an adequate response to any human-caused change that may or may not emerge.

Policymakers should resist pressure from lobby groups to silence scientists who question the authority of IPCC to claim to speak for "climate science." The distinguished British biologist Conrad Waddington wrote in 1941,

It is … important that scientists must be ready for their pet theories to turn out to be wrong. Science as a whole certainly cannot allow its judgment about facts to be distorted by ideas of what ought to be true, or what one may hope to be true (Waddington, 1941).

This prescient statement merits careful examination by those who continue to assert the fashionable belief, in the face of strong empirical evidence to the contrary, that human CO_2 emissions are going to cause dangerous global warming.

Reference

Waddington, C.H. 1941. *The Scientific Attitude*. London, UK: Penguin Books.

About the Authors

Dr. Craig D. Idso is founder and chairman of the Center for the Study of Carbon Dioxide and Global Change. Since 1998, he has been the editor and chief contributor to the online magazine CO2 Science. He is the author of several books, including *The Many Benefits of Atmospheric CO2 Enrichment* (2011) and *CO2, Global Warming and Coral Reefs* (2009). He earned a Ph.D. in geography from Arizona State University, where he lectured in meteorology and was a faculty researcher in the Office of Climatology.

Dr. Robert M. Carter is a stratigrapher and marine geologist with degrees from the University of Otago (New Zealand) and University of Cambridge (England). He is the author of *Climate: The Counter Consensus* (2010) and *Taxing Air: Facts and Fallacies about Climate Change* (2013). Carter's professional service includes terms as head of the Geology Department, James Cook University, chairman of the Earth Sciences Panel of the Australian Research Council, chairman of the national Marine Science and Technologies Committee, and director of the Australian Office of the Ocean Drilling Program. He is currently an emeritus fellow of the Institute of Public Affairs (Melbourne).

Dr. S. Fred Singer is one of the most distinguished atmospheric physicists in the United States. He established and served as the first director of the U.S. Weather Satellite Service, no part of the National oceanic and Atmospheric Administration (NOAA), and earned a U.S. Department of Commerce Gold Medal Award for his technical leadership He is coauthor, with Dennis T. Avery, of *Unstoppable Global Warming Every 1,500 Years* (2007, second ed. 2008) and many other books. Dr. Singer served as professor of environmental sciences at the University of Virginia, Charlottesville, VA (1971–94) and is founder and chairman of the nonprofit Science and Environmental Policy Project. He earned a Ph.D. in physics from Princeton University.

About NIPCC

The Nongovernmental International Panel on Climate Change (NIPCC) is what its name suggests: an international panel of nongovernment scientists and scholars who have come together to understand the causes and consequences of climate change. Because we are not predisposed to believe climate change is caused by human greenhouse gas emissions, we are able to look at evidence the Intergovernmental Panel on Climate Change (IPCC) ignores. Because we do not work for any governments, we are not biased toward the assumption that greater government activity is necessary.

The NIPCC traces its roots to a meeting in Milan in 2003 organized by the Science and Environmental Policy Project (SEPP), a nonprofit research and education organization based in Arlington, Virginia. SEPP, in turn, was founded in 1990 by Dr. S. Fred Singer, an atmospheric physicist, and incorporated in 1992 following Dr. Singer's retirement from the University of Virginia. NIPCC is currently a joint project of SEPP, The Heartland Institute, and the Center for the Study of Carbon Dioxide and Global Change.

NIPCC has produced seven reports to date:

Nature, Not Human Activity, Rules the Climate
Climate Change Reconsidered: The 2009 Report of the Nongovernmental International Panel on Climate Change (NIPCC)
Climate Change Reconsidered: 2011 Interim Report
Climate Change Reconsidered II: Physical Science
Climate Change Reconsidered II: Biological Impacts
Scientific Critique of IPCC's 2013 'Summary for Policymakers'
Why Scientists Disagree About Global Warming

These publications and more information about NIPCC are available at www.climatechangereconsidered.org.

About The Heartland Institute

The Heartland Institute is a national nonprofit research and education organization based in Chicago, Illinois. We are a publicly supported charitable organization and tax exempt under Section 501(c)(3) of the Internal Revenue Code.

Heartland is approximately 5,500 men and women funding a nonprofit research and education organization devoted to discovering, developing, and promoting free-market solutions to social and economic problems. We believe ideas matter, and the most important idea in human history is freedom.

Heartland has a full-time staff of 35. Joseph Bast is cofounder, president, and CEO. Dr. Herbert Walberg is chairman of the 10-member Board of Directors. Approximately 250 academics participate in the peer review of its publications and more than 175 elected officials pay annual dues to serve on its Legislative Forum.

Heartland has a long and distinguished history of defending freedom. We are widely regarded as a leading voice in national and international debates over budgets and taxes, environmental protection, health care, school reform, and constitutional reform. Five centers at The Heartland Institute conduct original research to find new ways to solve problems, turn good ideas into practical proposals for policy change, and then effectively promote those proposals to policymakers and the public.

For more information, visit our website at www.heartland.org, call 312/377-4000, or visit us at at 3939 North Wilke Road, Arlington Heights, Illinois.